T0282619

RHENIUM

DVI-MANGANESE,
THE ELEMENT OF ATOMIC NUMBER 75

RHENIUM

DVI-MANGANESE, THE ELEMENT OF ATOMIC NUMBER 75

BY

J. G. F. DRUCE

M.A., M.Sc. (Lond.), R. Nat. Dr. (Prague), F.R.I.C.
Fellow of the Chemical Society
Member of the Masaryk Academy of Work
Corresponding member of the Royal Bohemian Scientific Society
Member of the Netherlands Chemical Society, etc.

CAMBRIDGE
AT THE UNIVERSITY PRESS
1948

CAMBRIDGE
UNIVERSITY PRESS

University Printing House, Cambridge CB2 8BS, United Kingdom

Published in the United States of America by Cambridge University Press, New York

Cambridge University Press is part of the University of Cambridge.

It furthers the University's mission by disseminating knowledge in the pursuit of education, learning and research at the highest international levels of excellence.

www.cambridge.org
Information on this title: www.cambridge.org/9781107693241

First published 1948
First paperback edition 2014

A catalogue record for this publication is available from the British Library

ISBN 978-1-107-69324-1 Paperback

CONTENTS

PREFACE

The discovery of the element of atomic number 75, Mendeléef's dvi-manganese, may be said to have been made in 1925 when three announcements claiming the isolation of this congener of manganese were made almost simultaneously. Credit for the discovery is usually given to the German investigators, W. Noddack, I. Tacke and O. Berg, who called the new element rhenium, after Germany's 'Westmark', the Rhineland, just as they called eka-manganese (which they also claimed to have discovered) by the name of masurium, after Germany's 'Ostmark', Masurenland.

Some share in the honours of discovery is, however, surely due to the British co-discoverers and also to the Czechs, J. Heyrovský and V. Dolejšek. From 1922 onwards, F. H. Loring (together with the author) had been engaged upon a search for missing elements, including the possibility of a third missing congener of manganese, heavier than uranium and possessing an atomic number 93. Loring died in 1944 from injuries sustained in an air-raid.

Heyrovský is best known through his introduction of a new instrument for chemical researches of many kinds, namely, the polarograph. One of the first discoveries to be made with the polarograph was the presence of traces of dvi-manganese in crude manganese preparations. Dolejšek, who carried out the X-ray spectrographic examination of Heyrovský's preparations (and also of some of the author's), died at Terezín concentration camp early in 1945.

Nearly four hundred scientific communications dealing with rhenium have now appeared, and within a few years of its discovery four short works were published on rhenium, namely,

Rhenium (and Masurium), by P. W. Tyler, U.S. Bureau of Mines, 1931, pp. 17.

Das Rhenium, by Dr W. Schrotter, Stuttgart, 1932, pp. 59.

Renij, by Dr E. S. Kronman, Moscow, 1932, pp. 86.

Das Rhenium, by I. and W. Noddack, Leipzig, 1933, pp. 86.

Since 1933 much new knowledge has accumulated and the chemistry of rhenium has been fully investigated. The element, though still scarce, has become an article of commerce, and some of its salts are also obtainable commercially.

The present work is therefore an attempt to give a comprehensive survey of the chemistry of dvi-manganese, or rhenium, based upon the researches recorded in many journals published in many lands.

It emphasizes once again that progress in science is due to international effort.

J. G. F. DRUCE

LONDON, S.W. 16

February 1947

CHAPTER I

INTRODUCTION

SEARCH FOR THE CONGENERS OF MANGANESE

AFTER Mendeléef had enunciated the periodic law in 1869, attention was directed towards a search for missing elements which the Russian savant confidently predicted would ultimately be discovered to fill the blank spaces that he had left when drawing up his Table of the Elements. The fulfilment of his predictions to the minutest details, by the discovery of

Eka-boron	Scandium	(Nilson, 1875)
Eka-aluminium	Gallium	(Lecoq de Boisbaudran, 1879)
Eka-silicon	Germanium	(Winkler, 1886)

convinced men of science that the periodic law was a useful generalization, accounting for the relationships of the elements. It stimulated further investigations, but whereas Mendeléef had given full details concerning properties to be expected of the three elements cited above, his other prognostications were less specific. To the missing elements in group VII he gave the provisional names eka-manganese and dvi-manganese, but no detailed predictions of the properties to be expected of these elements. He merely assumed that they would resemble manganese in general properties.

Among the claims to the discovery of alleged elements that might conceivably have corresponded with eka-manganese are

Davyum	at.wt. 100	(Kern, 1877)
Ilmenium	at.wt. 104·6	(Hermann, 1846)
Lucium	at.wt. 100	(Barrière, 1896)
Nipponium	at.wt. 100	(Ogawa, 1908)

whilst possible precursors of dvi-manganese (rhenium) are

Uralium	at.wt. 187	(Guyard, 1869)
Pluranium	—	(Osann, 1828)
'Ruthenium'	—	(Osann, 1828)

It is interesting to note the davyum showed a thiocyanate (sulphocyanide) reaction identical with the one now used for rhenium. Rang (1), in a Periodic Table issued in 1895, placed davyum in the space for eka-manganese and uralium in that for dvi-manganese.

No new element appears to have been suspected or detected in manganese minerals prior to 1925, but both platinum ores and columbite (Noddacks's original sources of rhenium and masurium) have been fertile fields for alleged discoveries of new elements. In platinum ores Osann (2) claimed to have found three new ones, pluranium, polinium and ruthenium (not identical with the recognized ruthenium discovered by Claus in 1845). Osann later admitted that polinium was impure iridium and that 'pluranium oxide' was probably a mixture of the oxides of titanium, zirconium and silicon. His ruthenium, however, gave volatile reddish prisms which Berzelius pronounced to be new. Such a red body might have been impure rhenium trioxide, although in 1926 Zvjaginstsev (3) and other Russian workers recently re-examined platinum minerals and concluded that there could be no new element there, contrary to the claim of the Noddacks. It was Claus who, in the middle of last century, established the existence of the six recognized platinum metals, and since then no discovery of any new element in platinum ores has been substantiated. Several chemists have, however, found what they believed to be evidence of the presence of new bodies in platinum minerals. The most significant of these are uralium, announced by Guyard (4) in 1869, and davyum, claimed by Kern (5) from platinum residues. The stated density (20·25) and atomic weight (187) of uralium (or ouralium as it was sometimes transcribed from Russian) are almost identical with those of rhenium, but its malleability, ductility and its general resemblance to platinum in chemical properties indicate that it could not have been rhenium. Rose's pelopium (6) from Bavarian tantalite might likewise have been one of the congeners of manganese, but the so-called pelopic acid was subsequently regarded as impure niobic acid, and the claim that a new element had been found was dropped.

Of more significance was ilmenium, an element said to occur in ilmenite by Hermann (7). It was stated to have an atomic weight of 104·6. The density of the red-brown oxide, to which the formula IlO_2 was assigned, was 3·95–4·20, and it gave a characteristic red colour with potassium ferrocyanide. Salts of ilmenous and ilmenic acids, a white chloride and some double chlorides were also described. In the *Bulletin* of the Société Impériale de Moscou for 1872 Hermann gave a further account of the isolation of tantalum, niobium and ilmenium from columbite by fractional crystallization of the double potassium fluorides and sodium salts. He claimed to have isolated the metal by heating its double fluoride with metallic potassium. The black powder so obtained readily ignited in air, on warming, to give a pentoxide. Five years later (8) he brought forward a mass of evidence (dry tests, analyses, preparations, etc.) to show the elementary nature of ilmenium and to refute the criticisms of Marignac and others. Its isolation as a black powder that readily ignited in air would be consistent with the behaviour of rhenium. Little attention appears to have been given to Hermann's claims which he probably weakened by announcing another new element, neptunium (not to be confused with the artificially prepared element of atomic number 93), which he said was present in a columbite from Haddam, Connecticut.

The alleged rare earth element, lucium, patented by Barrière (9) might have corresponded with eka-manganese, but it was shown by Crookes (10) and by Shapleigh (11) to be impure yttrium, although it was accepted as a distinct element by several contemporary French savants.

The announcement in 1908 by Ogawa (12) that he had found a new body, nipponium, in thorianite, molybdenite and reinite, with an atomic weight of 100, again suggested that the eka-manganese gap had been at last filled. The reported chemical properties of nipponium accorded with those expected for a congener of manganese. Thus, it formed both acidic and basic oxides and an anhydrous chloride which gave a green solution and a green body on fusion with alkalis. About the same time Miss Evans (13) obtained what were thought to be indications of a new element in thorianite, and it was thought that this one was identical with nipponium. The claims for a new element did not, however, commend themselves to contemporary men of science, although Skrobal and Artmann concluded (**14**) from its reactions that it must be a new substance. The suggestion was also made that Boucher's 'new element' in cast iron (**15**) and one claimed by Ruddock in certain steels (**16**) might also be identical with Ogawa's nipponium. The reactions given by Boucher correspond in part with those for rhenium, but at the time it was considered (**17**) that they could largely be accounted for by the presence of molybdenum.

From the foregoing it will be observed that among the unrecognized elements there are several possible precursors of masurium, but only the very doubtful uralium and the still more improbable pluranium are possible precursors of rhenium. Renewed impetus was given to the search for the missing elements after Moseley had discovered that each element could be assigned an atomic number calculated from the frequency of the main X-ray spectral lines, since the square root of the frequency of vibration of the X-rays is proportional to the nuclear charge and therefore to the atomic number (**18**).

History of Rhenium

The discovery of dvi-manganese was first claimed by Noddack, Tacke and Berg in June 1925 (**19**). This was followed a few months later by the announcements of Druce and Loring (**20**) and of Heyrovský and Dolejšek (**21**).

From theoretical reasoning the German investigators argued that platinum ores and also certain minerals, notably columbite, should contain the missing congeners of manganese. They estimated that these elements would be present in native platinum to the extent of 10^{-3} to 10^{-4} and in columbite to the extent of 10^{-5} to 10^{-6}. From these sources they obtained a 'residue' containing 0·5 % of eka-manganese and 5 % of dvi-manganese. They named the new elements masurium and rhenium after the east and west German provinces of Masurenland and Rheinland, and supposed that there was about 10^{-13} part of eka-manganese and 10^{-12} part of dvi-manganese in the earth's crust, compared with 7×10^{-12} part of manganese and 10^{-2} part of iron.

Their claims to the discovery of these elements did not pass unchallenged, and for some time other investigators retained Mendeléef's provisional names. Whether the early criticisms of the work of I. and W. Noddack (Fräulein Ida Tacke married Dr Walter Noddack in 1926) are eventually substantiated or not, quantities of the element of atomic number 75 have since been isolated and the name rhenium has become the recognized one for this element.

Simultaneously, Loring and Druce were examining pyrolusite and crude manganese compounds for indications of the existence of an element beyond uranium and having an atomic number 93. In the course of this work (22) it was found that these sources contained traces of a new element which was identified as dvi-manganese. It was characterized by some chemical tests (formation of a sodium salt, precipitates with ferro- and ferricyanides, green spectral lines on a platinum wire flame test and a fugitive higher chloride) and by X-ray analysis (23). The results of the X-ray analysis were confirmed later by Prof. Manne Siegbahn (24) who made a spectroscopical examination of the original preparations and found lines characteristic for an element of atomic number 75 as Loring had previously announced (20).

The line $\lambda = 1{\cdot}43$ (1·4298) corresponded with L_{α_1} of element 75
$= 1{\cdot}4408$,, ,, L_{α_2} ,,
$= 1{\cdot}2358$,, ,, L_{β_1} ,,
$= 1{\cdot}204$,, ,, L_{β_2} ,,

Heyrovský and Dolejšek also showed the presence of dvi-manganese in commercial (and even 'pure') manganese salts. They detected the new element during investigations on the electrolytic deposition potentials of manganese salt solutions using Heyrovský's newly introduced polarographic dropping mercury cathode method. The presence of one part of the new element in 20,000 parts of the preparations was corroborated spectroscopically and a re-examination of both sets of preparations from manganese sources confirmed previous announcements (25). Later, however, Heyrovský has stated (26) that his polarographic results with bivalent manganese compounds may not have been conclusive.

In 1926, W. Prandtl (27) criticized all claims to the discovery of a new element. In particular he denied the occurrence of dvi-manganese in columbite, a mineral with which he had had considerable experience. Again, as stated above, some Russian men of science (Zvjaginstsev and Seljakov and Korsunski) were also unable to detect the presence of dvi-manganese in native platinum, and they estimated that it could not be present in amounts exceeding 0·0003 % (3). Since they made this announcement no reference has been made to the occurrence of rhenium in platinum minerals. Both columbite and platinum ores have been abandoned by the German investigators as sources of rhenium, which is now extracted from molybdenum glance, a source indicated by Loring (28), and was obtained in considerable quantity from an unspecified industrial sulphide residue by a process introduced by Feit (29). Should a great demand for columbite in making certain stainless steels again arise there might be an accumulation of residues that could be examined further to ascertain the presence (or absence) of rhenium in this mineral.

Whilst defending themselves against the criticisms of their own work, W. and I. Noddack (30) joined with Prandtl in his criticisms of other work. They cited that of Hartley and Ramage (31) and of Bosanquet and Keeley (32), in which these investigators stated that they were unable to find any congener of manganese in pyrolusite by spectroscopic methods. They considered that manganese minerals are among the poorest sources of rhenium. Loring (33), however, directed attention to the weakness of their own spectroscopic evidence and indicated the precautions taken in his work. Actually he obtained the X-ray spectral lines of rhenium with a sample (Kahlbaum's) of commercial manganese phosphate (34).

The original method of I. and W. Noddack for isolating crude eka- and dvi-manganese from platinum ores consisted in treating the mineral with *aqua regia*, evaporating the solution and igniting the residue, which was then reduced in hydrogen. The part insoluble in *aqua regia* was heated in a stream of chlorine and the chloride reduced with zinc. The two products were together heated alternately in hydrogen and oxygen and gave sublimates of the oxides of osmium, ruthenium and arsenic, together with a new body, darkened by hydrogen sulphide. In the case of columbite this was first fused with sodium hydroxide and nitrate to remove the bulk of the iron, niobium and tantalum, and the filtrate was treated repeatedly with hydrogen sulphide in both alkaline and acid solution. The redissolved sulphide solutions were concentrated to small bulk and precipitated with mercurous nitrate. The combined precipitates were reduced and the final product, on X-ray examination, was stated to show the presence of both congeners of manganese.

In the process outlined by their patent (35) I. and W. Noddack simplified the procedure somewhat. In this patent it is stated that molybdenum minerals form suitable raw materials for the extraction of rhenium. The mineral containing this element is dissolved as far as possible in nitric acid and molybdenum removed by addition of phosphoric acid and ammonium nitrate. The rhenium (together with other elements) is then precipitated with hydrogen sulphide. The sulphide precipitate is converted into crude oxide, and by repeating the process and finally by fractional sublimation, a product rich in rhenium was obtained.

In directing attention to the difficulties met with in preparing pure manganese chloride from residues left after making chlorine from pyrolusite, W. Smith (36) stated in 1924 that, in quantitative experiments, some 0·04 % of material was unaccounted for. This suggested a possible source for the manganese congeners. The first method adopted for extracting dvi-manganese (rhenium) from the manganese salts of commerce was as follows:

Quantities of 100 g. of crude manganese sulphate or chloride were dissolved in about 600 c.c. of water together with 50 g. of ammonium chloride and 25 c.c. of 0·880 ammonia. Hydrogen sulphide was passed through the solution until it was saturated with the gas. After standing, the precipitate was filtered off. It should have contained any iron sulphide and aluminium and chromium hydroxides if these metals were present in the original salt, together with manganese sulphide. The filtrate was retreated with hydrogen sulphide until no further precipitation occurred. Filtrate and washings were allowed to stand for several days whilst still alkaline and saturated with hydrogen sulphide to make sure that as much manganese as possible had been removed. The final filtrate was afterwards boiled for some minutes to expel excess of hydrogen sulphide and some ammonia. It was neutralized with hydrochloric acid and evaporated to dryness and the residue was heated to drive off ammonium compounds and to remove any sulphur set free during the treatment. The residue was dissolved in dilute acetic acid, and a slight excess of ammonium oxalate was added to precipitate the calcium present. The filtrate from the calcium oxalate precipitate was evaporated to dryness and again heated to drive off ammonium salts. The 75 mg. of residue so obtained from 400 g. of manganese sulphate contained about 1 % of dvi-manganese according to estimates made at the time. No further enrichment could be achieved by repeating the process with accumulated preparations.

The extraction and concentration of the new element from pyrolusite (which contains very little of element of atomic number 75) by these methods proved slow and unsatisfactory. It was ascertained that the dvi-manganese formed a volatile compound which was evolved in the excess of chlorine when pyrolusite was acted upon by hydrochloric acid alone (37). The chlorine containing traces of dvi-manganese compounds was either (*a*) absorbed by alkalis, (*b*) precipitated with barium hydroxide, or (*c*) acted upon by metallic zinc which readily converted it into crude hydrated rhenium hydroxide. The products from (*c*) and (*b*) were readily converted into rhenium heptoxide by heating in a stream of oxygen. From solutions of (*a*) it was possible to crystallize out some potassium perrhenate.

Still another method consisted in heating the pyrolusite or other source of dvi-manganese with concentrated nitric acid and then working up the solution which contained perrhenic acid. By the last two methods sufficient quantities of the new element were obtained to enable some of its properties and reactions to be investigated and some of its compounds to be prepared and studied.

After Heyrovský had detected rhenium in 'pure' manganese sulphate by means of his dropping mercury cathode, he separated a product rich in the new element by dipping strips of platinum and zinc into concentrated manganese sulphate solutions. The X-ray spectrum of this product gave at least three clear lines characteristic of the element of atomic number 75. The lines Dolejšek observed were

$$L_{\alpha_1},\ 1{\cdot}430;\quad L_{\beta_1},\ 1{\cdot}233;\quad L_{\beta_2},\ 1{\cdot}2043;\quad \text{and } L_{\gamma_1},\ 1{\cdot}059.$$

Later Heyrovský and Dolejšek showed (38) that the new element accumulated in manganese amalgam when this and platinum foil were placed in concentrated manganese sulphate solution. They have also had what might be indications of the presence of eka-manganese in some of the enrichment products.

The technical preparation of rhenium from waste sulphide slime in an unspecified metallurgical process was worked out by Feit (29) in 1930. The weathered slime was extracted with water to give a brown-green solution containing chiefly copper and nickel. By regulated addition of amounts of ammonia most of the copper and nickel and some zinc were separated as double sulphates. Further additions of ammonia to the thick black mother liquor yielded dark crystals of the ammonium salts of molybdic, vanadic and phosphoric acids and their complexes. The pale yellow filtrate yielded crystals of crude potassium perrhenate on addition of potassium chloride in excess.

The impure grey product was washed with cold water, dissolved in hot dilute sodium hydroxide, filtered and cooled to deposit purer potassium perrhenate, and two more recrystallizations gave a pure product.

Prof. E. S. Kronman (39) and his colleagues at the Moscow Rare Metals Institute have extracted rhenium compounds (equivalent to about 5 mg. of rhenium per kg. of material) from Russian and Japanese molybdenum glance. With this product they were able to make several important researches, especially on the microchemical reactions of rhenium.

In extracting rhenium from molybdenum glance the mineral was treated with nitric acid and then with sulphuric acid. The rhenium was distilled off in the volatile fraction either with a current of air at 300° C. for 3 hr. or equally well in a stream of hydrochloric acid gas at 200° C.

According to Driggs (40) when ores or concentrates containing rhenium are strongly heated with sodium bisulphate and nitrate, rhenium heptoxide volatilizes. It is purified by conversion into the sulphide (with arsenic sulphide). This, on reduction with hydrogen and heated to about 1000° C. to drive off arsenic, leaves rhenium alone.

REFERENCES

(**1**) Rang. *Chem. News*, 1895, **72**, 217.
(**2**) Osann. *Pogg. Ann.* 1829, **15**, 158.
(**3**) Zvjaginstsev. *Nature, Lond.*, 1926, **117**, 262.
Seljakov and Korsunski. *Phys. Z.* 1927, **28**, 478.
(**4**) Guyard. *Chem. News*, 1879, **40**, 57.
(**5**) Kern. *Chem. News*, 1877, **36**, 4.
(**6**) Rose. *Pogg. Ann.* 1846, **69**, 115.
(**7**) Hermann. *J. Prakt. Chem.* 1846, **38**, 91.
(**8**) *Bulletin de la Société Impériale de Moscou*, 1877, [2], **15**, 105.
(**9**) Barrière. *Chem. News*, 1894, **74**, 159 and 212.
(**10**) Crookes. *Chem. News*, 1894, **74**, 259.
(**11**) Shapleigh. *J. Franklin Inst.* 1897, **114**, 68.
(**12**) Ogawa. *J. Coll. Sci. Tokio*, 1908, **15**, 1.
(**13**) Evans (Miss). *J. Chem. Soc.* 1908, **93**, 666.
(**14**) Skrobal and Artmann. *Chem. Ztg*, 1909, **33**, 143.
(**15**) Boucher. *Chem. News*, 1897, **86**, 99 and 182.
(**16**) Ruddock. *Chem. News*, 1897, **86**, 118.
(**17**) Jones. *Chem. News*, 1897, **86**, 171.
(**18**) Moseley. *Phil. Mag.* 1913, **26**, 1024.
(**19**) Noddack, Tacke and Berg. *Naturwissenschaften*, 1925, **13**, 567.
(**20**) Loring and Druce. *Chem. News*, 1925, **131**, 273 and 337.
(**21**) Heyrovský and Dolejšek. *Nature, Lond.*, 1925, **116**, 782.
(**22**) Loring's first paper on missing elements appeared in *Chem. News* in 1922.
(**23**) Loring. *Nature, Lond.*, 1926, **117**, 153.
(**24**) *Chem. Weekbl.* 1932, **29**, 57.
(**25**) Dolejšek, Druce and Heyrovský. *Nature, Lond.*, 1926, **117**, 159.
(**26**) Heyrovský. *Nature, Lond.*, 1935, **133**, 670.
(**27**) Prandtl. *Z. angew. Chem.* 1926, **39**, 1049.
(**28**) Loring. *Chem. News*, 1926, **133**, 356.

(**29**) Feit. *Z. angew. Chem.* 1930, **43**, 459.
(**30**) W. and I. Noddack. *Metallbörse*, 1926, **16**, 2129.
(**31**) Hartley and Ramage. *Trans. Chem. Soc.* 1897, **71**, 533.
(**32**) Bosanquet and Keeley. *Phil. Mag.* 1924, **48**, 145.
(**33**) Loring. *Nature, Lond.*, 1926, **117**, 622; *Chem. News*, **133**, 276.
(**34**) Loring. *Nature, Lond.*, 1926, **117**, 448.
(**35**) D.R.P. 483,495 and British Patent, 317,035, 28 July 1930.
(**36**) Smith. *Chem. News*, 1924, **128**, 1.
(**37**) Druce. *Chem. News*, 1931, **143**, 66.
(**38**) Heyrovský and Dolejšek. *Chem. Listy*, 1926, **20**, 4.
(**39**) Kronman, Bibikova and Aksenova. *J. Appl. Chem.*, Russia, 1934, **7**, 47.
(**40**) U.S. Patent, 1,911,934, 30 May 1933.

CHAPTER II

THE ISOLATION AND PROPERTIES OF RHENIUM

FROM theoretical reasoning, the German investigators, Noddack, Tacke and Berg, argued that platinum ores and also certain minerals, notably columbite, should contain eka- and dvi-manganese. They estimated that the manganese 'homologues' would be present in native platinum to the extent of 10^{-3} to 10^{-4} and in columbite to the extent of 10^{-5} to 10^{-6}. From these sources, as described above, they obtained a 'residue' alleged to contain 0·5 % of eka-manganese and 5 % of dvi-manganese. From various preparations and enriched products attempts were then made to isolate, first pure compounds and then the elements themselves. No progress was made with eka-manganese, and it has not been isolated by ordinary chemical means. More success has been achieved with dvi-manganese, or rhenium, and several methods are now available for its isolation.

PREPARATION OF ELEMENTARY RHENIUM

The following methods have been used for the preparation of metallic, or elementary, rhenium from its compounds:

(*a*) Reduction of the heated oxides in a stream of hydrogen. The higher oxides are quickly reduced to the dioxide which, at higher temperatures (above 500° C.), is reduced to the element.

(*b*) Reduction of heated rhenium sulphides with hydrogen. With the hepta-sulphide some sulphur is disengaged but it is soon volatilized from the element by the stream of hydrogen.

(*c*) Reduction of the halides and oxyhalides. Thus, Van Arkel (1) acted upon rhenium trichloride with tungsten at high temperatures and fused the resulting alloy of rhenium and tungsten into wire. Hot hydrogen also reduces rhenium halides at 250–300° C.,

$$2ReCl_3 + 3H_2 = 2Re + 6HCl.$$

(*d*) By electrolysis of aqueous alkaline solutions (German Patent, 723,303).

(*e*) Reduction of potassium or ammonium perrhenate with hydrogen at 400° C. The ammonium salt yields a purer and also more active product, since the rhenium from the potassium salt occludes some alkali, even after prolonged washing,

$$2KReO_4 + 7H_2 = 2KOH + 2Re + 6H_2O.$$

This is the most usual method and may be carried out by heating the perrhenate in a platinum boat within a silica tube in a stream of purified hydrogen.

(*f*) By the thermite reaction with rhenium dioxide,

$$3ReO_2 + 4Al = 2Al_2O_3 + 3Re.$$

(*g*) Rhenium can be conveniently recovered from the nitron perrhenate precipitate obtained in quantitative estimations of the element (see p. 22). The precipitate is heated in a stream of hydrogen until it melts and decomposes. The swollen mass is then extracted with alcohol before a final heating to redness in hydrogen when fairly pure powdered rhenium remains.

PROPERTIES OF RHENIUM

Elementary rhenium is obtained by the methods described as a brown-black powder, having an apparent density of 10·4. This powder can be compressed and electrically fused into rods of a silver white, resistant metal which does not tarnish readily in air and which is more difficult to attack even than the powder. The metal rods are exceptionally hard. The density of rhenium in the form of rods is 21·04, and they melt, according to Becker and Moers (2), at 3440 ± 50° abs. The coefficient of expansion is 0·00000467 perpendicular to the *c*-axis. More recently, Jaeger and Rosenbaum (3) have given the melting-point as about 3160° C. and the specific heat at 0–20° C. as 0·03262. Agte and his co-workers (4) showed that rods of rhenium are ductile in the cold, and when hot they can be forged, hammered and rolled. The tensile strength is 50·6 kg./sq.m. with an elongation of 20 %.

The temperature dependence of resistance from 0·85 to 373° abs. was studied by Aschermann and Justi (5) for rhenium rod and plate, specially made from sintered powder at 1000 and 2400° C. respectively. The results indicated a temperature coefficient of 0·46 % over the range 0–100° C. At about 0·95° abs. rhenium is super-conductive.

The electrode potential of rhenium against a normal calomel electrode with twice normal sulphuric acid is 0·6 V., so that rhenium occupies a position in the electrochemical series between copper and thallium.

The mean values for the parachor of rhenium were determined by Briscoe, Robinson and Rudge (6), who obtained the values

Derived from	ReO_3Cl	Re_2O_7	ReO_2Cl_3
Parachor is	76·4	68·9	78·9

(Compared with W = 90 and Os = 75).

Rhenium amalgamates with mercury and alloys with tungsten and other elements. One eutectic with tungsten at 2892° C. corresponds with WRe, and another at 2822° C. corresponds with WRe_2. A compound W_2Re_3 may also exist. With phosphorus rhenium begins to react at 750° C. Tensimetric and X-ray examination (7) point to the existence of compounds having the formulae ReP_3, ReP_2 and ReP.

Rhenium silicide, $ReSi_2$, isomorphous with molybdenum silicide, $MoSi_2$, has been reported (8).

The element is usually obtained as a dark, loose powder, and in this form it readily ignites in air, usually giving the volatile white heptoxide,

$$4Re + 7O_2 = 2Re_2O_7.$$

It also combines directly with chlorine to give the trichloride, or the less stable pentachloride. It has been widely stated (9) that rhenium tetrachloride and possibly a hexachloride are also produced when rhenium combines with chlorine under special circumstances. The main product is, however, the trichloride, though when rhenium reacts with fluorine the principal product is the hexafluoride.

Rhenium is attacked by nitric acid which ultimately converts it into perrhenic acid. Sulphuric acid has but little action, even at elevated temperatures. What action there is, above 200° C., would appear to be due to sulphide formation, accompanied by the evolution of sulphur dioxide. Under ordinary circumstances hydrochloric acid has no appreciable action upon rhenium.

In the finely divided form rhenium is attacked by fused alkalis, especially in the presence of suitable oxidizing agents such as potassium nitrate, sodium peroxide or even atmospheric oxygen. The ultimate product is a perrhenate or a higher oxide of rhenium.

Direct union also takes place between rhenium and sulphur and between rhenium and selenium (**10**).

The element acts as a catalyst, like nickel, in the hydrogenation of unsaturated hydrocarbons. Thus, it readily converts equimolecular mixtures of ethylene and hydrogen into ethane at 300–400° C. It has also been claimed by Tropsch and Kassler (**11**) that the element, either alone or in conjunction with copper, acts as a catalyst in the conversion of carbon monoxide and hydrogen into methane. At higher temperatures, however, the carbon monoxide is quantitatively decomposed into carbon and the dioxide, and there is some formation of metal carbide.

As a catalyst for the reduction of nitric oxide at 200–400° C. rhenium produced nitrogen but not ammonia (**12**). The same Russian authorities have used the element, deposited on porcelain, for the dehydrogenation of *n*-propyl alcohol and formation of propaldehyde. They have also effected similar oxidations of other alcohols. The optimum temperature was 400° C. and the yields were sometimes almost quantitative.

Purified rhenium disulphide has been used as a catalyst for the conversion of primary alcohols into the corresponding aldehydes and of isopropyl alcohol into acetone. Similarly, cyclohexanol at 550° C. gave a 78·8 % yield of phenol (with traces of benzene), but this catalyst did not show any tendency to reduce carbon monoxide (**13**).

When free rhenium is absorbed on to charcoal, either alone or with other catalysts, it has been used in the production of phenols and cresols according to a German Patent, 693,707 (1940).

Colloidal Rhenium

Colloidal rhenium can be obtained (**14**) from potassium rheni-chloride, or potassium perrhenate in hydrochloric acid, in the presence of gum arabic, by reduction on warming with 1 % hydrazine and formaldehyde. The colloid is purified by dialysis. As might be expected, it catalyses the decomposition of hydrogen peroxide, the synthesis of ammonia and the hydrogenation of maleic and cinnamic acids.

The Optical Spectrum of Rhenium

In 1931 W. F. Meggers (**15**) determined the arc spectrum of rhenium (using potassium perrhenate) in the region from 2300 A. in the ultra-violet to 8800 A. in the infra-red. About 2000 lines were recorded. The following table by Loring (**16**) gives the principal lines in Ångström units together with their relative intensities:

Wave-length	Intensity	Wave-length	Intensity
2887·66	60	4522·71	100
2992·35	50	4791·42	80
2999·39	80	**4889·15**	2000
3067·38	50	4923·93	80
3399·31	60	5270·98	400
3424·61	120	5275·57	1000
3451·88	200	5834·31	300
3460·46	600	6307·70	75
3464·46	400	6321·88	100
3725·76	50	6350·73	50
4136·45	100	6605·17	60
4144·36	60	6813·37	100
4222·46	200	6829·91	120
4257·61	50	6971·48	60
4358·69	50	7024·10	50
4394·37	80	7640·92	80
4513·31	300	7912·90	50
4516·62	50		

The strongest line, 4889·15, is undoubtedly the most sensitive one for the identification of this element. This blue line and the green ones, 5270·98 and 5275·57, are the most brilliant ones in the visible spectrum of rhenium.

Borovik and Gudris found (**17**) that the intensity of the lines 3451·88, 3460·46 and 3464·72 is unaffected by as much as 30 %

of molybdenum but manganese interferes. These lines serve to detect as little as 0·002 % of rhenium in, for example, calcite or molybdenite.

Using infra-red plates Meggers explored the long-wave portions of the arc spectrum of rhenium from 7200 to 11,000 A. and recorded many new lines (**18**). S. Piña de Rubies and J. Dorronsova (**19**) also measured some 200 lines in the region of 2320 to 2500 A. Schober and Birke (**20**) compared the rhenium lines with certain lines in the solar spectrum, but, as Loring pointed out, the identity of rhenium in the sun has not been established. Its presence has, however, been shown in meteorites from South Africa, Russia and the U.S.A.

The X-ray Spectrum of Rhenium

The following table gives the calculated and observed values in Ångström units for the principal X-ray spectrum lines of rhenium according to various investigators:

	Calculated		Found			
Line	Loring	Wennerlöf	Loring	Dolejšek	Berg and Tacke	Beuthe
L_{α_1}	1·4298	1·4298	1·430 1·428	1·430	1·429	1·4298
L_{α_2}	1·4408	—	1·4408	—	1·4407	1·4407
L_{α_3}	—	—	—	—	—	1·4240
L_{β_1}	1·2358	1·23604	1·233	1·2353	1·2352	1·2339
L_{β_2}	—	1·20408	—	1·2043	1·2048	1·2038
L_{β_3}	—	—	—	—	1·215	—
L_{γ_1}	—	—	—	1·059	—	1·0589
L_{γ_2}	—	—	—	—	—	1·0299
L_{γ_3}	—	—	—	—	—	1·0236

Induced Radioactivity of Rhenium

When rhenium powder was bombarded with slow neutrons from beryllium, with deuterium (heavy hydrogen) or with fast neutrons from lithium it showed activities with half-life periods of about 76 and 90 hr. It has also been established that rhenium has the following isotopes (**21**):

184	52 ± 2 days	187	stable
185	stable	188	16 (or 20) hr.
186	93 hr.		

Valency of Rhenium

Since rhenium occupies the position of dvi-manganese in group VII of Mendeléef's Periodic Table its maximum valency should be 7. Actually this is its most characteristic valency and is shown in several of its most important and most stable compounds, such as the heptoxide, Re_2O_7, the perrhenic acid, $HReO_4$, and its salts, including that of potassium, $KReO_4$.

In compounds like the hexafluoride, ReF_6, the trioxide, ReO_3, the oxytetrachloride, $ReOCl_4$, and the oxythiocyanate, $ReO(CNS)_4$, the valency of rhenium is 6. A valency of 5 is shown by the element in a few compounds like the pentachloride, $ReCl_5$; whilst in rhenium dioxide, tetrafluoride and the double halides of the type K_2ReCl_6, the valency is 4.

The existence of the trihalides, like the trichloride, $ReCl_3$, and double salts like $RbReCl_4$, as well as trimethylrhenium, $Re(CH_3)_3$, indicates that the element can also show a valency of 3. On the other hand, it has been shown that in glacial acetic acid the trichloride has a molecular weight corresponding with the formula Re_2Cl_6, and Geilmann and Wrigge represent the compound as being built up of two rhenium tetrachloride complexes in which two chlorine atoms are shared (22):

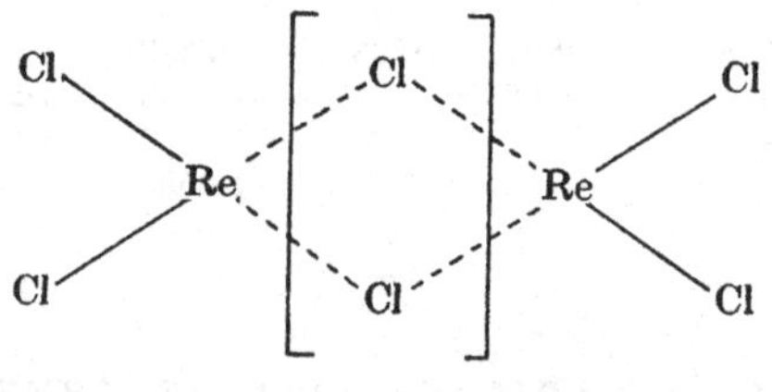

It might have been expected that, like manganese, rhenium would not form any univalent compounds, though bivalent ones, analogous to the chloride and sulphate of manganese, could be reasonably sought. I. and W. Noddack stated that solutions of univalent and bivalent rhenium compounds were obtained during the reduction of solutions of tervalent derivatives, but no definite compounds were reported. It has since been claimed that oxides of both univalent and bivalent rhenium can be obtained by reducing perrhenic acid with zinc and cadmium respectively in the presence of hydrochloric acid (23). When reduced with the aid of the Jones' reductor method (zinc amalgam) perrhenic acid behaves analogously to the halogen oxy-acids. Since these are known to be reduced to the hydracids (periodates give iodides of hydriodic acid), it has been inferred that hydrorhenic acid, HRe, is formed, although no such compound has yet been actually isolated.

Tests for Rhenium

Among the early qualitative tests used for detecting rhenium was a borax bead test. Rhenium imparts a black colour to a borax bead heated in the reducing flame. Yahoda (24) has modified this test by using a sodium carbonate bead in which the black colour first formed can be converted into a yellow one in the oxidizing flame. As little as 0·015 mg. is found to give this test.

In 1931 Geilmann and Wrigge (25) made an exhaustive study of the dry reactions of rhenium compounds. Among other things they found that potassium and silver perrhenates can be fused without perceptible decomposition. Organic perrhenates decompose leaving more or less carbon and lower valency rhenium compounds. Ammonium perrhenate gives off some ammonia and forms a white sublimate containing rhenium, whilst a dark residue (? Re) is left. Heated so that air has access, a variety of products are noticeable. Odourless fumes colouring the flame green are given off. White drops, that are probably impure rhenium heptoxide, or perrhenic acid, appear on the coolest part of the tube. Below them is a liquid and then the very hygroscopic crystals of purer rhenium heptoxide. Nearer to the substance being heated (ammonium perrhenate) is a blue product or deposit. This is probably the oxide, Re_3O_8, and may be obtained by burning a little sulphur in the tube with the rhenium compound. If strongly heated in air the blue substance goes to the heptoxide.

On charcoal, in the oxidizing flame, rhenium compounds give off plenty of fumes. A green colour is developed and a bronze-red deposit is formed. In the reducing flame (with or without sodium carbonate) the free element is liberated.

Borax beads are dark in the reducing flame; in the oxidizing flame the colour gradually goes as the rhenium is converted into perrhenate.

Certain colour reactions observed with rhenium compounds have been perfected by Tougarinoff (26). Thus, through interaction of perrhenates with potassium ferrocyanide in the presence of stannous chloride, a brown coloration is observed. With alcoholic dimethylglyoxime solution instead of ferrocyanide, a yellow to red coloration develops and a green fluorescence is observed when the mixture is warmed; as little as 0·01 mg. of rhenium gives these reactions.

Geilmann, Wrigge and Weibke (27) were the first to introduce a useful test whereby rhenium is detected by means of thiocyanate and stannous chloride. A yellow to orange colour is produced when potassium thiocyanate is added to a hydrochloric acid solution of rhenium in the presence of stannous chloride. The coloured substance is soluble in ether and remains as a red solid when the solvent evaporates (see p. 68).

The reduction of tellurates by stannous chloride is catalysed by the presence of perrhenates, and this has been utilized (28) to detect as little as $2{\cdot}5 \times 10^{-8}$ g. of the element.

Characteristic crystalline precipitates were obtained by Kronman and Berkman (29) on mixing a drop of perrhenate solution reduced with hydriodic acid, with solutions of potassium iodide, rubidium iodide, caesium chloride, thallium nitrate or sulphate and mercurous nitrate. Molybdates and tungstates must be absent as they interfere with the test. These precipitates of double halides resemble potassium platinichloride in appearance and can be formed conveniently on microscope slides.

Rhenium can be separated from molybdenum and colorimetrically estimated by reduction with mercury after previous oxidation with permanganate (30). The perrhenate solution, containing molybdate, in strong (1 : 1) hydrochloric acid is shaken with ether. The acid layer contains most rhenium and a little molybdenum, whilst the ether layer contains most molybdenum with only traces of rhenium. The acid fraction is diluted to 3 % hydrochloric acid and treated with potassium thiocyanate and mercury and is then extracted again with ether. The ether layer is now free from rhenium which is practically all in the dilute acid layer. To it is added potassium thiocyanate and stannous chloride, and the mixture is once more extracted with ether and the extract examined colorimetrically for rhenium by comparison with standard preparations.

The procedure has been employed even when 10 mg. of molybdenum were present to less than 1 mg. of rhenium. The raw material is treated with 10 mg. of iron, as ferric chloride, and a few drops of potassium permanganate. A slight excess of ammonium hydroxide is next added, and the mixture is warmed on a steam bath to ensure the formation of perrhenate. The residue after evaporation is dissolved in hydrochloric acid (25 c.c.) and 2 c.c. of 20 % potassium thiocyanate solution are added, together with 20 c.c. of ether. Mercury is run in from a small separating funnel and the whole is shaken. Extraction of the acid layer is repeated with 15 c.c. of ether and 1 c.c. of thiocyanate; then with 1 c.c. of a stannous chloride solution (made from 350 g. of stannous chloride crystals in 200 c.c. of 1 : 1 hydrochloric acid and the whole made up to 1 l.). The last ether extract is coloured and is compared for rhenium content with standards in Nessler tubes, diluting if necessary.

Another selective reagent for rhenium is toluene-3:4-dithiol, which gives a green complex with rhenium (but also with molybdenum) and a blue one with tungsten (**31**). Wenger and Duckert, after examining various methods for detecting rhenium, concluded that many were not satisfactory. Under the microscope they favoured the potassium or caesium halide test. Spot tests with stannous chloride and sodium tellurate or dimethylglyoxime were considered satisfactory and reliable (**32**).

Quantitative Estimation of Rhenium

Heyrovský (**33**) estimated the dvi-manganese (rhenium) content of 'pure' manganese solutions and of his enriched products by the nature of the 'wave' at $-0{\cdot}98$ V. (from the zero of the calomel electrode). The intensity of the *L* series of lines of the X-ray spectra were also used at first as affording some indication of the amount of rhenium present in the crude preparations. Spectroscopic examination alone is of service in detecting rhenium in minerals on account of the very small amount of the element present even in the richest ores. Piña de Rubies (34) used it for estimating the element by observing the number of lines corresponding with a known rhenium content.

In 1930 Geilmann and Voigt (35) elaborated a procedure for determining rhenium as perrhenic acid by means of nitron (the organic base 1:4-diphenyl-3:5-endo-anilino-4:5-dihydro-1:2:4-triazole, used similarly for estimating nitrates with which it forms an insoluble precipitate). Nitron perrhenate is practically insoluble, since 100 c.c. of water dissolve only 0·018 mg. of the salt. This gravimetric method affords one of the best means for estimating rhenium in perrhenates, and in such compounds that can be converted into perrhenic acid or its salts. It has the advantage that the rhenium can be recovered from the nitron perrhenate precipitate (see p. 13).

Krauss and Steinfeld (36) made use of the insolubility of thallium perrhenate as a means of estimating the element which, as in Geilmann and Voigt's method, must first be converted into the perrhenic acid or one of its soluble salts. The solution, acidified with 20 % acetic acid solution, is precipitated with thallium acetate. The precipitate of thallium perrhenate, $TlReO_4$, is filtered off, washed, drained and dried at 140° C.; halogens interfere with this estimation.

Briscoe, Robinson and Stoddart (37) showed that hydrated rhenium dioxide, ReO_2, $2H_2O$, was quantitatively precipitated from perrhenate solutions by the action of zinc and hydrochloric acid, and they used this as a means of estimating the element. Later, Geilmann and Hurd have asserted that the product so formed may not be pure and that it does not conform constantly to the above formula (38).

Solutions containing tervalent and quadrivalent rhenium can be estimated by oxidation with potassium permanganate or ferric sulphate solution. Thus, potassium rhenichloride, K_2ReCl_6, solutions require three equivalents of the standard oxidizing agents, whilst the green solutions of tervalent rhenium require four for complete oxidation.

Another volumetric method that has been proposed is to oxidize the rhenium compound with, for example, hydrogen peroxide to perrhenic acid and to titrate this with a standard alkali solution. Instead of titrating the acid it may be precipitated as potassium, or thallium (or even nitron) perrhenate and estimated gravimetrically. Indeed, Tollert proposed (39) the use of perrhenic acid for the estimation of potassium in view of the fact that the perrhenate is one of the least soluble of potassium salts.

In estimating the oxides of rhenium, Geilmann and Hurd pointed out that rhenium heptoxide could not be titrated with reducing agents, but when dissolved in water it could be titrated with alkalis using any indicator (all that they tried were successful), since perrhenic acid is strongly dissociated. Lower oxides, e.g. ReO_3, ReO_2 and Re_2O_3, are best estimated by oxidation with ferric sulphate and subsequent titration of the ferrous salt so formed.

Three Russian chemists, Michajlova, Pevsner and Archipova (40), have suggested a method for the microdetermination of rhenium. The element is first precipitated (along with molybdenum) as sulphide and oxidized with alkaline hydrogen peroxide. Excess of this reagent is removed, the solution is then neutralized with ammonia and molybdenum separated by adding 8-hydroxy-quinoline in the presence of acetic acid and ammonium acetate. Rhenium is finally precipitated as nitron perrhenate, $C_{20}H_{16}N_4, HReO_4$, containing 0·4442 part of rhenium.

An electrolytic method for the estimation of rhenium was devised by Tomíček in 1939 (41). The procedure consists in a preliminary cathodic reduction of the perrhenate ion in 5% sulphuric acid solution containing about 1–20 mg. of rhenium. The solution is maintained at 70° C. for the electrolysis, which takes about 3 hr. to complete. The anode is of platinum gauze and the cathode is a platinum wire. Current density is 0·25 amp./sq.dm., and the potential is 2·34 V. The rhenium deposited on the cathode is determined by oxidizing with hydrogen peroxide and subsequent titration with centinormal sodium hydroxide solution using methyl red or bromo-cresol purple as indicator.

Another electrolytic method (42) involves the separation of rhenium from ammoniacal solutions and is claimed to be more rapid.

The most recent method for the determination of rhenium from molybdenum minerals is that described by Hiskey and Meloche (43). About 4 g. of the mineral are dissolved in 20 c.c. of concentrated nitric acid together with 5 c.c. of the fuming acid. The mixture is heated to near the boiling-point and then 50 c.c. of hydrochloric acid are added and the mixture is slowly evaporated. During this process the residue is not allowed to become dry, for, from time to time, more hydrochloric acid is added (120–150 c.c. in all). When the volume is finally reduced to about 25 c.c. the mixture is cooled and 75 c.c. of concentrated sulphuric acid are added and the whole transferred to a distilling flask where it is distilled at 260–270° C. with a current of steam (two parts) and carbon dioxide or air (one part). When the distillate amounts to 250 c.c. (having been cooled in ice), after about 2½ hr., bromine dissolved in potassium bromide solution is added and the colour is compared with standards containing 10, 50 and 100 μg. of rhenium, after adding 100 c.c. of hydrochloric acid, 10 c.c. of a 20 % sodium thiocyanate solution and 10 c.c. of a 20 % stannous chloride solution. The comparison is made in Nessler tubes.

The polarographic method has been extended to examine rhenium compounds by Lingane (44). The perrhenate solution in 2–4 N-hydrochloric, or perchloric, acid as supporting electrolyte is reduced to quadrivalent rhenium at the dropping mercury cathode. In 4 N-perchloric acid the diffusion current is well defined and is proportional to the concentration of perrhenate ion. The half-wave potential is −0·4 V. In 2 N-hydrochloric acid the half-wave potential is −0·45 V. and in 4·2 N-acid it is −0·31 V. In neutral, unbuffered solutions of potassium chloride a double wave is observed. The first is due to the reduction to the rhenide ion, Re^-, and the second is due to the discharge of hydrogen. In phosphate-buffered solutions of pH 7 the perrhenate ion gives a catalytic wave at −1·6 V.

Polarograms of reduced solutions in 1–2 N-sulphuric acid solution at 0° C. show three anodic waves whose half-wave potentials are $\alpha = -0{\cdot}54$, $\beta = -0{\cdot}34$ and $\gamma = -0{\cdot}07$ V. respectively to the calomel electrode. A similar polarogram is obtained in normal perchloric acid except that the β wave separated into two ($-0{\cdot}42$ and $-0{\cdot}26$ V.) and the γ wave shifted to about 0·1 V. more positive than in the sulphuric acid solution. The α wave represented oxidation to Re^{++}, the β wave to Re^{+++} and the γ wave to Re^{v} or Re^{vii} (i.e. perrhenate).

The Atomic Weight of Rhenium

From its position in the Periodic Table the atomic weight of rhenium must lie between those of tungsten, 184, and osmium, 190. In 1926, Loring suggested (45) the value 187 as the approximate atomic weight of the new element, and Washburn arrived at the high figure of 188·4 by a modification of Harkins and Williams's schemes for calculating atomic weights (46).

According to Jaeger and Rosenbaum (47) the specific heat of the element is 0·0346, and on calculating the atomic weight according to Dulong and Petit's law the value of 185 is obtained.

From analyses of the disulphide, I. and W. Noddack (48) obtained the figure $188{\cdot}71 \pm 0{\cdot}15$. The oxides and chlorides of rhenium are unsuitable for atomic weight determinations, but the stable disulphide could be obtained free from heptasulphide and free sulphur by heating in a stream of carbon dioxide at 900° C. This disulphide, on strongly heating in hydrogen, was reduced to rhenium.

A later and more reliable determination by Hönigschmid and Sachtleben (49) was based upon the conversion of silver perrhenate into bromide, these authorities having found that a long time was needed to obtain pure rhenium disulphide, whilst its final reduction was incomplete even after heating in hydrogen for some hours. They therefore made highly purified silver perrhenate in three different ways and fused their specimens to ensure that they were perfectly anhydrous.

The three methods used to make the very pure silver perrhenate were:

(1) Addition of concentrated potassium perrhenate solution to excess of silver nitrate solution. The silver perrhenate

precipitated was recrystallized from silver nitrate solution to remove any occluded potassium salt.

(2) Addition of silver nitrate solution (made from pure silver dissolved in pure nitric acid) to perrhenic acid solution. The silver perrhenate was recrystallized from the purest hot water.

(3) Silver oxide was dissolved in hot perrhenic acid and the filtrate cooled in ice.

All three specimens were considered to be highly pure, especially after fusion. Altogether 51·82860 g. of silver perrhenate gave 27·17309 g. of silver bromide, so that the atomic weight of rhenium was 186·31 ± 0·02.

Aston (50) has shown that rhenium has two isotopes, 185 and 187, in such a proportion that the atomic weight should be 186·22.

The Occurrence and Abundance of Rhenium

In their search for suitable sources of rhenium I. and W. Noddack examined no less than 1800 minerals and meteors (51). From among the many substances they examined, the following include the most promising sources of the element:

Mineral	Origin	Composition	Rhenium content parts per million
Alvite	Norway	$(Zr, Hf)SiO_4$	0·6
Gadolinite	Norway	$(Fe, Be)_2Y_2Si_2O_{10}$	0·03
Thortveitite	Norway	$Y_2Si_2O_7$	0·6
Columbite	Norway	$(Fe, Mn)Nb_2O_6$	0·2
Tantalite	South Africa	$Mn(Ta, Nb)_2O_6$	0·03
Wolframite	Czechoslovakia	$(Fe, Mn)WO_4$	0·02
Rutile	Austria	TiO_2	0·02
Copper glance	Montana, U.S.A.	Cu_2S	0·04
Copper pyrites (bright)	Mansfeld, Germany	Cu_3FeS_3	0·02
Copper pyrites	Clausthal, Germany	$CuFeS_2$	0·08
'Kupferschiefer'	Mansfeld, Germany	Earthy Cu pyrites	0·03
Molybdenum glance	Norway	MoS_2	21·0
Molybdenum glance	Japan	MoS_2	10·0
Molybdenum glance	Siberia	MoS_2	0·6
Molybdenum glance	Colorado, U.S.A.	MoS_2	1·8
Iron pyrites	Norway	FeS_2	0·01
Silver sulphide	Germany	Ag_2S	0·07
Platinum ores	Urals		0·8
Meteoric iron	South Africa		0·01
Meteoric iron	Mexico		0·004
Meteoric iron	Augustinovka, Russia		0·008

The element has also been detected in Finnish gadolinite by Aartovaara (**52**), who states that it is present in this mineral in greater quantity than in any other so far examined. Otherwise, it will be observed that the richest source is molybdenum glance. The likely association of rhenium with molybdenum was suggested by Loring in 1926 (see p. 6). Molybdenite from northern Wisconsin has also been reported to possess a high rhenium content (**53**).

REFERENCES

(**1**) Van Arkel. *Metallwirtschaft*, 1934, **13**, 405.
(**2**) Becker and Moers. *Metallwirtschaft*, 1830, **9**, 1063.
(**3**) Jaeger and Rosenbaum. *Proc. K. Akad. Wet. Amst.* 1933, **36**, 786.
(**4**) Agte *et al.* *Naturwissenschaften*, 1931, **19**, 108.
(**5**) Aschermann and Justi. *Phys. Z.* 1942, **43**, 207.
(**6**) Briscoe, Robinson and Rudge. *J. Chem. Soc.* 1932, p. 2673.
(**7**) Haraldsen. *Z. anorg. Chem.* 1935, **221**, 398.
(**8**) Wallbaum. *Metallkunde*, 1942, **33**, 378.
(**9**) Briscoe, Robinson and Rudge. *J. Chem. Soc.* 1931, p. 2263.
Yost and Shull. *J. Amer. Chem. Soc.* 1932, **54**, 4657.
Druce. *Chem. Weekbl.* 1926, **23**, 318.
(**10**) Briscoe, Robinson and Stoddart. *J. Chem. Soc.* 1931, p. 1439.
(**11**) Tropsch and Kassler. *Zpr. Úst. Věd. Výzk. Uhlí*, 1932, **2**, 13.
(**12**) Platonov *et al.* *Ber. dtsch. chem. Ges.* 1935, **68**, 761.
(**13**) *J. Gen. Chem.*, Russia, 1941, **11**, 683.
(**14**) Zenghelis and Stathis. *C.R. Acad. Sci., Paris*, 1939, **209**, 797.
(**15**) Meggers. *J. Franklin Inst.* 1931, **211**, 373.
(**16**) Loring. *Chem. News*, 1931, **142**, 321.
(**17**) Borovik and Gudris. *J. Appl. Chem.*, Russia, 1936, **9**, 937.
(**18**) Meggers. *U.S. Bur. Stand. J. Res.* 1933, **10**, 757.
(**19**) Piña de Rubies and Dorronsova. *An. Soc. esp. Fís. Quím.* 1931, **31**, 412.
(**20**) Schober and Birke. *Naturwissenschaften*, 1931, **19**, 211.
(**21**) Fajans and Sullivan. *Phys. Rev.* 1940, **58**, 276.
(**22**) Geilmann and Wrigge. *Z. anorg. Chem.* 1935, **223**, 144.
(**23**) Young and Irvine. *J. Amer. Chem. Soc.* 1937, **59**, 2648.
(**24**) Yahoda. *Industr. Engng Chem.* (Anal. ed.), 1936, **8**, 133.
(**25**) Geilmann and Wrigge. *Z. anorg. Chem.* 1931, **199**, 65.
(**26**) Tougarinoff. *Bull. Soc. chim. Belg.* 1934, **43**, 111.
(**27**) Geilmann, Wrigge and Weibke. *Z. anorg. Chem.* 1932, **208**, 217.
(**28**) Anisimov. *J. Appl. Chem.*, Russia, 1944, **17**, 658.
(**29**) Kronman and Berkman. *Z. anorg. Chem.* 1933, **211**, 277.
(**30**) Hoffman and Lundell. *Bur. Stand. J. Res.* 1939, **23**, 479.
(**31**) Miller. *Analyst*, 1944, **69**, 112.
(**32**) Wenger and Duckert. *Helv. chim. Acta*, 1942, **25**, 599.

(**33**) Heyrovský. *Chem. Listy*, 1926, **20**, 4.
(**34**) Piña de Rubies. *An. Soc. esp. Fís. Quím.* 1932, **30**, 918.
(**35**) Geilmann and Voigt. *Z. anorg. Chem.* 1930, **193**, 666.
(**36**) Krauss and Steinfeld. *Z. anorg. Chem.* 1931, **197**, 52.
(**37**) Briscoe, Robinson and Stoddart. *J. Chem. Soc.* 1931, p. 666.
(**38**) Geilmann and Hurd. *Z. anorg. Chem.* 1933, **214**, 260.
(**39**) Tollert. *Naturwissenschaften*, 1930, **18**, 849.
(**40**) Michajlova, Pevsner and Archipova. *Z. anal. Chem.* 1932, **91**, 25.
(**41**) Tomíček. *Trans. Electrochem. Soc.* 1939, **76**, 197.
(**42**) Voigt. *Z. anorg. Chem.* 1942, **248**, 225.
(**43**) Hiskey and Meloche. *Industr. Engng Chem.* (Anal. ed.), 1940, **12**, 503.
(**44**) Lingane. *J. Amer. Chem. Soc.* 1942, **64**, 1001.
(**45**) Loring. *Chem. News*, 1926, **132**, 407.
(**46**) Washburn. *J. Amer. Chem. Soc.* 1925, **48**, 2351.
(**47**) Jaeger and Rosenbaum. *Proc. K. Akad. Wet. Amst.* 1933, **36**, 786.
(**48**) Noddack, I. and W. *Electrochemie*, 1928, **34**, 629.
(**49**) Hönigschmid and Sachtleben. *Z. anorg. Chem.* 1930, **191**, 309.
(**50**) Aston. *Proc. Roy. Soc.* A, 1931, **132**, 487.
(**51**) Noddack, I. and W. *Z. phys. Chem.* 1931, A, **154**, 207.
(**52**) Aartovaara. *Tekn. Fören. Finl. Förh.* 1932, **52**, 157.
(**53**) Works, Mrs L. P. *Rocks & Miner.* 1941, **16**, 92.

CHAPTER III

THE OXIDES OF RHENIUM

THE following oxides of rhenium have been reported:

ReO_4, or Re_2O_8	Rhenium tetroxide, or octoxide
Re_2O_7	Rhenium heptoxide
ReO_3	Rhenium trioxide
Re_3O_8	—
Re_2O_5	Rhenium pentoxide
ReO_2	Rhenium dioxide
Re_2O_3	Rhenium sesquioxide
ReO, or Re_2O_2	Rhenium monoxide
Re_2O	Rhenium suboxide

The heptoxide and the dioxide are the best known, whilst the existence of two (the tetroxide and the pentoxide) is doubtful. The higher oxides are acidic, and it might have been expected that the lower ones, like those of manganese, would show basic properties. The dioxide is not basic but the lowest oxides probably are, although at present very little is known about them.

?*Rhenium tetroxide*, ReO_4 or Re_2O_8. When rhenium or its dioxide is oxidized at temperatures not exceeding 150° C. in a rapid stream of oxygen, a volatile white solid oxide is obtained. From analytical data and from its reactions (and also from the fact that the more common heptoxide was at first supposed to be a deep yellow body), I. and W. Noddack and others (1) concluded that this white substance must be the tetroxide (or octoxide if the double formula is assigned to it). Its existence was questioned by Briscoe, Robinson and Rudge (2), who considered the heptoxide to be the highest oxide of rhenium. They regarded the alleged tetroxide as a modification of the heptoxide, or as that oxide which had deliquesced. This view is most probably correct. Hagen and Sieverts (3) also failed to obtain oxygen from the alleged octoxide when it was heated either alone or in a stream of carbon dioxide. It was expected that if decomposition occurred oxygen would be readily detected,

$$2Re_2O_8 = 2Re_2O_7 + O_2.$$

The oxide has been said to react with water to form perrhenic acid and apparently hydrogen peroxide as well, since although no oxygen was evolved the resulting solution was said to decolorize potassium permanganate with the evolution of a little oxygen,

$$Re_2O_8 + 2H_2O = 2HReO_4 + H_2O_2.$$

With potassium hydroxide the octoxide reacted to form the perrhenate but no oxygen was observed.

As the highest valency of rhenium is 7, the following formula has been suggested for the octoxide (assuming that it does exist):

```
O                  O
  \\              //
O=Re—O—O—Re=O
  //              \\
O                  O
```

The compound is said to be easily converted into the heptoxide by heat and is both unstable and deliquescent. Sulphur dioxide reduced it to a violet-coloured substance. When hydrogen sulphide was led over it a white deposit of sulphur was formed and then dark rhenium heptasulphide made its appearance,

$$Re_2O_8 + 8H_2S = Re_2S_7 + 8H_2O + S.$$

Heated gently in a stream of hydrogen it is soon reduced to the dioxide. I. and W. Noddack assigned to it a density of 8·4, and a melting-point of 150–155° C., but nothing has been reported concerning it for some years and its existence is very doubtful.

Rhenium heptoxide, Re_2O_7. This light yellow solid is obtained when rhenium or its lower oxides are heated at above 160° C. in air or oxygen. It is formed most readily at just below red heat. Its heat of formation is 2·95 kg.cal. (4).

Rhenium heptoxide is a volatile solid, very deliquescent, readily combining with water to form perrhenic acid. It melts at about 304° C. I. and W. Noddack give the melting-point as 220° C., but this must be too low for a pure product. It can be sublimed in dry oxygen or nitrogen unchanged, at temperatures above 300° C., in the form of thin platelets. Its molecular weight was found (I. and W. Noddack) to be 502 at 520° C., using Victor Meyer's method. The theoretical value for Re_2O_7 is 490.

It is reduced by carbon monoxide and by sulphur dioxide to

lower oxides. With hydrogen at 300° C. it gives rhenium dioxide, whilst at 800°C. it is completely reduced to the element. Rhenium heptoxide dissolves in alcohol, acetone and methyl alcohol, but not in dry ether or in carbon tetrachloride.

Rhenium trioxide, ReO_3. The existence of a trioxide of rhenium was early suspected, since an unstable yellow-red solution was obtained as an intermediate product when nitric acid acted on rhenium or on the dioxide. It is also produced during the incomplete combustion of the element, or certain of its compounds, in air, and may also be present in the unstable intermediate products of the reduction of perrhenic acid and of rhenium heptoxide. It was first definitely isolated by Biltz and Lehrer (5), who heated together rhenium heptoxide and the finely divided element in absence of air at 200–250° C.,

$$3Re_2O_7 + Re = 7ReO_3.$$

Biltz later (6) made the trioxide by heating 5 g. of the dioxide with 10 g. of heptoxide at 300° C. for a week and then rubbing the product with more heptoxide and reheating a further three days. Excess of the heptoxide was sublimed away,

$$ReO_2 + Re_2O_7 = 3ReO_3.$$

Its heat of formation is given as $82{\cdot}5 \pm 10$ kg.cal. It is weakly paramagnetic and its electrical resistance at ordinary temperatures is 2×10^{-3}; in liquid air this falls to 1×10^{-4} (7).

Rhenium trioxide is a red microcrystalline solid with a metallic glance. Its density by the pyknometer method is 6·9 and by the X-ray spectral method 7·4. It does not appear to be polymeric. It readily unites with more oxygen forming the heptoxide. When heated alone at 400° C. *in vacuo* it breaks up into heptoxide and dioxide. With excess of potassium hydroxide solution it forms the unstable rhenate and the perrhenate, whilst with fused sodium oxide, perrhenate and rhenite are said to result,

$$2Na_2O + 3ReO_3 = 2NaReO_4 + Na_2ReO_3.$$

The trioxide liberated iodine from potassium iodide solution and was reduced by stannous chloride with the formation of an indefinite black substance.

?*Rhenium pentoxide*, Re_2O_5. Briscoe, Robinson and Rudge (2) suggested that the red substance, to which Biltz and Lehrer later assigned the formula ReO_3, was probably the pentoxide. They described this compound as a purplish red crystalline powder which resulted when the heptoxide was heated in a sealed pyrex tube with rhenium,

$$4Re + 5Re_2O_7 = 7Re_2O_5.$$

Analytical data appeared to support their views.

The purplish red oxide, which showed a 'green streak' and was bottle green in thin flakes, was stable in dry air and could be heated in oxygen up to 300° C., or with sulphur up to 190° C. without change. Above 300° C. *in vacuo* it distilled, leaving a black residue of indefinite composition. There were white, yellow and violet films, or sublimates. Dry hydrogen chloride had no action, nor had hydrochloric acid solutions, but chlorine gave a green-yellow vapour, leaving a brown solid—probably the oxytetrachloride. The oxide was insoluble in dilute or concentrated hydrochloric or sulphuric acids, and in aqueous potassium hydroxide. The substance was attacked by warm nitric acid (even dilute) and by fused caustic alkalis.

Rhenium dioxide, ReO_2. When higher oxides of rhenium are partly reduced by hot hydrogen the dioxide is left. It is perhaps best obtained (6) by heating together rhenium and the heptoxide for a day at 600–650° C.,

$$3Re + 2Re_2O_7 = 7ReO_2.$$

More or less pure hydrated dioxide is formed when perrhenic acid or perrhenate solutions are reduced with zinc and hydrochloric acid (8),

$$HReO_4 + 3H = ReO_2 + 2H_2O.$$

Also, when the acid is evaporated with hydrazine, rhenium dioxide results. Electrolytic reduction of perrhenic acid and its salts constitute a still further means of preparing this hydrated dioxide, and it is also a product of the hydrolysis of rhenichlorides (see p. 56).

When heated *in vacuo* to 650–680° C. the hydrated oxide loses the two molecules of water of constitution. This dioxide was first prepared by Briscoe and his co-researchers. Its constancy and purity have been called into question by German workers, possibly on account of the tendency it shows to appear first as a colloidal precipitate. The passage of carbon dioxide into such a colloidal suspension renders the hydrated dioxide more easily filterable. It dries to constant weight in a desiccator over magnesium perchlorate (anhydrone) (8). The water of constitution can be driven off by heating under reduced pressure.

Anhydrous rhenium dioxide is a very dark brown (almost black) solid, which sometimes assumes a bluish tinge. Its density is 11·4 and it is weakly paramagnetic, whereas manganese dioxide is strongly so. Its specific resistance is 8×10^{-4} at room temperature and 2×10^{-4} at the temperature of liquid air (7).

When strongly heated in a stream of hydrogen it is reduced to the element which may remain in a pyrophoric condition, re-igniting if brought into the air whilst still warm. At elevated temperatures (above 750° C.) in absence of oxygen, and best in a vacuum, it breaks up into rhenium and the heptoxide, which volatilizes away,

$$7ReO_2 = 2Re_2O_7 + 3Re.$$

It is acted upon by hydrogen chloride gas, oxychlorides being formed. Rhenium tetrachloride has also been stated to be among the products of this reaction but no chlorine is evolved, since rhenium dioxide is not an oxidizing agent. Indeed, it is if anything a reducing agent, since quadrivalent rhenium compounds tend to oxidize so that the element is brought into the heptavalent state.

There is evidence that rhenium dioxide reacts with sulphur dioxide but no definite product has been isolated. When treated with such oxidizing agents as nitric acid, hydrogen peroxide, chlorine water or bromine water, rhenium dioxide is readily oxidized to perrhenic acid. When fused with alkalis in the presence of air, perrhenates are the ultimate products,

$$4ReO_2 + 4KOH + 3O_2 = 4KReO_4 + 2H_2O.$$

In the absence of air, rhenates and rhenites are formed, e.g.

$$ReO_2 + 2NaOH = Na_2ReO_3 + H_2O.$$

At 500° C. in an atmosphere of nitrogen, barium rhenate resulted from the fusion of rhenium dioxide with sodium hydroxide and barium hydroxide. Barium rhenate, $BaReO_4$, is a grass-green solid, whereas the perrhenate is colourless and barium manganate is bluish green.

Hyporhenites, e.g. Na_3ReO_4, are also produced when excess of alkali is used in the fusion. They are all unstable in moist air.

The formula Re_3O_8, has been assigned by I. and W. Noddack and others to the violet-blue reduction product of rhenium heptoxide formed when this oxide is acted upon by carbon monoxide or sulphur dioxide. The same compound has been observed during the oxidation of lower rhenium oxides.

Rhenium sesquioxide, Re_2O_3. Geilmann, Wrigge and Biltz (9) obtained this oxide in a hydrated form by hydrolysing rhenium trichloride and by acting upon it with alkalis in the absence of air. It is a black solid which rapidly oxidizes in air to perrhenic acid. Ferric sulphate and potassium permanganate solutions readily oxidize it in the same way. With alkalis it is reported to break up into bivalent, quadrivalent and heptavalent rhenium derivatives,

$$9Re''' \rightarrow 6Re'' + 2Re^{iv} + Re^{vii}.$$

When dried *in vacuo* the hydrated product is still apparently not pure Re_2O_3. It appears to decompose water, and analyses of the moist product from the hydrolysis of rhenium trichloride show a rhenium to oxygen ratio of from 2 : 3·2 to 2 : 3·9. In some respects rhenium sesquioxide resembles ferrous hydroxide.

Rhenium monoxide, ReO. Reduction of perrhenic acid with cadmium, using dilute hydrochloric acid and excluding air, gave a residue containing 83·20–85·85 % rhenium and also water. This dark amorphous substance is considered (10) to be the hydrated monoxide, ReO, H_2O, since it is not identical with the sesquioxide, and the analytical data accord best with this formula.

Rhenium suboxide, Re_2O. This oxide is stated to result in a yield of 25 % during the reduction of dilute perrhenic acid solutions with zinc and hydrochloric acid (10), air being excluded during the reaction. The perrhenic acid solution is made by dissolving 0·5–1·5 m.equiv. of the pure acid in 350 c.c. of 2N-hydrochloric acid solution. To this 100 m.equiv. of zinc (six pieces weighing 1 g.) were added. Periodically, 6 c.c. of concentrated hydrochloric acid were run in until no further effervescence occurred. Some rhenium still remained in solution. The black deposit was soluble in nitric acid and in bromine water, but was not readily attacked by alkaline chromate solutions, or by ferric sulphate. It contained only traces of zinc and halogen, and its composition agreed with the formula Re_2O, $2H_2O$.

REFERENCES

(**1**) See *Chem. News*, 1931, **143**, 66.
(**2**) Briscoe, Robinson and Rudge. *Nature, Lond.*, 1932, **129**, 618.
(**3**) Hagen and Sieverts. *Z. anorg. Chem.* 1932, **208**, 367.
(**4**) Roth and Becker. *Z. phys. Chem.* 1932, A, **159**, 27.
(**5**) Biltz and Lehrer. *Nachr. Ges. Wiss. Göttingen*, 1931, pp. 191–8.
(**6**) Biltz. *Z. anorg. Chem.* 1933, **214**, 225.
(**7**) Perakis, Kapatos and Kydriadides. *Praktika*, 1934, **8**, 163.
(**8**) Druce. *Chem. & Ind.* 1935, **54**, 54.
(**9**) Geilmann, Wrigge and Biltz. *Z. anorg. Chem.* 1933, **214**, 239.
(**10**) Young and Irving. *J. Amer. Chem. Soc.* 1937, **59**, 2648.

CHAPTER IV

PERRHENIC ACID AND ITS SALTS

UNLIKE permanganic acid, perrhenic acid is both colourless and stable. It resembles perchloric acid in some respects but is much more stable and is not an oxidizing agent, except under special circumstances.

Several methods are available for the preparation of perrhenic acid:

(*a*) It results from the direct union of the heptoxide with water,

$$Re_2O_7 + H_2O = 2HReO_4.$$

(*b*) Oxidation of the element with nitric acid yields a solution of perrhenic acid,

$$3Re + 7HNO_3 = 3HReO_4 + 3H_2O + 7NO.$$

(*c*) Nitric acid, hydrogen peroxide, chlorine water and many other oxidizing agents convert lower rhenium oxides into the acid,

$$2ReO_2 + 4H_2O + 3Cl_2 = 2HReO_4 + 6HCl.$$

The accompanying hydrochloric acid is driven off by heating.

(*d*) The oxidation of rhenium sulphide with nitric acid and other strong oxidizing agents also produces perrhenic acid. This constitutes a ready means for the recovery of the element in a convenient form.

The acid obtained by the methods outlined above is usually in dilute solution. When it is concentrated upon a water-bath it becomes more viscous, and finally a liquid is obtained which is about twice as heavy as water and contains over 60 % of the pure substance, $HReO_4$. This liquid resembles orthophosphoric acid in consistency but is yellowish in colour. The pure acid is, however, colourless and so are its salts, except those that owe their colour to the cation (e.g. copper perrhenate).

Perrhenic acid is a strong monobasic acid. It dissolves the metals magnesium, iron and zinc with evolution of hydrogen. It also acts upon many metal oxides, hydroxides (of Al, Zn and Fe) and carbonates with the formation of normal perrhenates:

$$Mg + 2HReO_4 = Mg(ReO_4)_2 + H_2,$$
$$Al(OH)_3 + 3HReO_4 = Al(ReO_4)_3 + 3H_2O,$$
$$CuCO_3 + 2HReO_4 = Cu(ReO_4)_2 + H_2O + CO_2.$$

It neutralizes alkalis and can be titrated by them with the use of the common indicators such as methyl red and phenolphthalein. The heat of neutralization has been found to be 13·5 kg.cal.

Perrhenic acid is very stable and not easily reduced. Neither the acid nor its salts exhibit any marked oxidizing powers. Some perrhenates are so stable that they can be fused and sometimes made red hot or even boiled without decomposition. One instance of oxidizing tendencies, however, has been observed, namely, its power to liberate bromine from hydrobromic acid.

Gaseous hydrogen bubbled through the acid has no action upon it. Sulphur dioxide causes the transitory appearance of a yellow colour (attributed to rhenic acid, H_2ReO_4), but this rapidly disappears on shaking the solution with air. Sulphuric acid is formed incidentally. Hypophosphorous acid behaves in a similar way towards perrhenic acid.

Sodium amalgam, stannous chloride and hydrazine all reduce the acid, first to the yellow substance, but a dark brown precipitate soon begins to appear as the final reduction product. When this precipitate is examined it is found to be a mixture of rhenium and its dioxide. The dioxide, in a hydrated condition, is obtained without admixed metal when perrhenic acid is reduced with zinc and hydrochloric acid. Sometimes this action shows a step-by-step reduction, the solution first becoming yellow and then greenish before the very dark brown precipitate of $ReO_2, 2H_2O$ appears. Magnesium with hydrochloric acid, or even acetic acid (1), gives this hydrated oxide from potassium perrhenate.

According to Hölemann (2) ferrous, stannous or titanous sulphates in concentrated sulphuric acid reduce potassium perrhenate to quinquevalent rhenium compounds, whilst chromous chloride in sulphuric acid reduces a perrhenate to quadrivalent rhenium derivatives.

The action of hydrogen sulphide on perrhenic acid differs according to the circumstances. When the gas is led into the dilute acid or its salts, thio-derivatives ($HReO_3S$, $KReO_3S$, see p. 63) are first formed (3). Rhenium sulphides are precipitated from strong solutions, especially if acidified with much hydrochloric acid (4). The best results are obtained when the solution contains about 10 % of hydrochloric acid.

Perrhenic acid catalyses the oxidation of ammonia to nitric acid by means of oxygen or hydrogen peroxide. It also gives a red colour with hydrogen peroxide alone, and Hagen and Sieverts (5) have attributed this colour to the presence of a perrhenic acid. According to Jakob and Jeżowska (6) perrhenic acid undergoes electro-reduction in hydrochloric acid of six to eight normal strength,

$$Re^{vii} \rightarrow Re^{v},$$

with poor current yield and marked cathodic polarization. The yellow-green salts, $K_2Re(OH)_2Cl_3$ and $(NH_4)_2Re(OH)_2Cl_3$, are described as arising when two equivalents of hydroiodic acid are added to a strong hydrochloric solution of potassium or ammonium perrhenate. These salts can be recrystallized from concentrated hydrochloric acid. They give an intense green compound, soluble in ether, with potassium thiocyanate. With alkalis it is stated that an unstable black hydrated oxide, $ReO(OH)_3$, H_2O, was precipitated, but this is the same as the hydrated dioxide.

It has been suggested by I. and W. Noddack (7) that the compound hitherto called perrhenic acid should be called metaperrhenic since a tribasic compound with more water, H_3ReO_5, has been obtained, together with certain of its salts, such as $Ba_3(ReO_5)_2$. This acid would thus be meso-perrhenic acid. The existence of series of salts such as

Meso-perrhenates	$3M_2O.Re_2O_7$	Yellow to red
Rhenates	$M_2O.ReO_3$	Green
Hyporhenates	$M_2O.Re_2O_5$	Yellow
Rhenites	$M_2O.ReO_2$	Brown

has been asserted, but the isolation of many of them has so far not been confirmed. The formation of sodium rhenite is referred to on p. 31.

Barium meso-perrhenate, $Ba_3(ReO_5)_2$, is stated by Scharnow (8) to be left as hexagonal prisms on evaporating a solution of barium perrhenate with excess of barium hydroxide under conditions excluding carbon dioxide which decomposes the meso-perrhenate, especially when moist. I. and W. Noddack, however, described (9) this compound as a lemon-yellow solid formed when barium perrhenate is acted upon with sodium hydroxide. They agree that the acid, H_3ReO_5, is weaker than carbonic acid and that water hydrolyses its salts. The hyporhenates are further subdivided into meta-, pyro- and ortho-salts, e.g.

$NaReO_3$	Sodium hyporhenate	(Meta-hyporhenate)
$Na_4Re_2O_7$	Sodium pyrorhenate	(pyro-hyporhenate)
Na_3ReO_4	Sodium ortho-hyporhenate	

The alkali rhenites, Na_2ReO_3 and K_2ReO_3, remain unchanged on heating in nitrogen up to 600° C. and are analogous to the manganites. Dilute acids act upon them slowly with the liberation of rhenium dioxide and the formation of the sodium or potassium salt of the acid used, but concentrated hydrochloric acid converts them into rhenichlorides or into the free chlororhenic acid,

$$K_2ReO_3 + 6HCl = K_2ReCl_6 + 3H_2O$$

or

$$K_2ReO_3 + 8HCl = H_2ReCl_6 + 2KCl + 3H_2O.$$

Among the salts of perrhenic acid that have been described are those of sodium, potassium, rubidium, caesium, ammonium, copper, silver, barium, neodymium, lanthanum, manganese and nickel, together with a whole series of ammine derivatives, as well as the salts of nitron and other organic bases. The potassium salt is described first because it is by far the most important.

Potassium perrhenate, $KReO_4$. The potassium salt of perrhenic acid is prepared by neutralizing hot perrhenic acid with potassium hydroxide or carbonate solution,

$$HReO_4 + KOH = KReO_4 + H_2O.$$

Fine crystals of the salt separate out as the solution cools. Sometimes the neutralization produces a micro-crystalline precipitate when the solubility of the salt is exceeded.

Potassium perrhenate also results when potassium chloride is added to perrhenic acid and the solution evaporated. Hydrochloric acid is evolved with the steam,

$$HReO_4 + KCl = KReO_4 + HCl.$$

According to Becker and Moers (**10**) the heat of formation of the salt is 263 kg.cal.

Potassium perrhenate forms small, anhydrous, tetragonal, bipyramidal crystals which can be fused at about 550° C. and even heated to the point of volatilization without decomposition. Vorländer and Dalichau (**11**) give the melting-point as 552 ± 3° C. (corr.), and they found that the boiling-point was 1370° C. at 763 mm. and 1359° C. at 752 mm. Above 600° C. the salt showed a definite vapour pressure and lost weight appreciably. A later determination of the melting-point by Kleese and Hölemann (**12**) gave the value 518° C. Its density is about 4·89. In toluene, however, Biltz found (**13**) the value 11·4.

Potassium perrhenate is even less soluble than the perchlorate. The table below is based upon unpublished determinations, together with those of Puškin and Kovać (**14**), of I. and W. Noddack (**15**) and of Kleese and Hölemann (the values above 100° C. are theirs and are given in percentages).

Solubility of potassium perrhenate

Temp. ° C.	gr. $KReO_4$ in 100 c.c. water	Temp. ° C.	% $KReO_4$ in 100 c.c. solution
2·01	0·4945	109	12·6
8·30	0·5207	112	14·0
10·20	0·5777	154	26·3
16·90	0·8350	194	39·7
23·80	1·1580	220	50·7
30·90	1·5410	239	59·9
35·00	1·7920	290	71·9
38·95	2·016	335	84·6
44·85	2·525	401	89·3
50·45	3·128	445	94·4
65·80	5·001	470	96·8
86·15	7·522	498	98·4
100·30	9·484	518	100·0 (m.p.)

The heat of solution has been determined by Roth and Becker (**16**) to be 13·93 kg.cal. at 16·7° C. and 13·80 kg.cal. at

21·7° C. ($\pm$ 3 kg.cal.). The presence of potassium hydroxide or chloride further depresses the solubility of potassium perrhenate and this assists in freeing it from chromates, molybdates, tantalates, osmiates and tungstates.

It is, however, more soluble in acid solutions, including that of perrhenic acid itself. In neutral or slightly acid solution potassium perrhenate on electrolysis gave rhenium and rhenium dioxide (hydrated) at the platinum or mercury cathodes. According to Hölemann (17) rhenium so liberated amalgamated with the mercury. In strongly acid solutions potassium perrhenate gave violet or green solutions on electrolysis.

Rubidium perrhenate, $RbReO_4$. This salt has been obtained in the form of colourless, octahedral pyramids in a similar manner to the potassium salt. It is somewhat more soluble than the latter which it closely resembles. Like rubidium perchlorate it can be coloured by crystallization with the permanganate.

Caesium perrhenate, $CsReO_4$. This salt apparently crystallizes in several forms and is even less soluble than the potassium salt.

Sodium perrhenate, $NaReO_4$. The sodium salt is prepared in the same ways as the potassium compound, especially by neutralizing perrhenic acid with sodium hydroxide, though heating is not necessary on account of the greater solubility of the sodium salt. Indeed, it is the most soluble of the perrhenates, for at ordinary temperatures 100 c.c. of water dissolve 100 g. of sodium perrhenate. It is rather deliquescent and can be melted at about 300° C., and it is not decomposed at 500° C. When liquefied it remains colourless.

Ammonium perrhenate, NH_4ReO_4. This salt is also obtained by neutralizing perrhenic acid with ammonium hydroxide. It is fairly soluble in water, and at ordinary temperatures its solubility is 6 and it crystallizes from saturated solutions in platelets. When heated it decomposes, giving a white sublimate (which contains both ammonia and rhenium, as heptoxide). A black residue, containing rhenium dioxide, remains behind. If ammonium perrhenate is heated in a stream of hydrogen a residue of rhenium is obtained; this is one of the best methods of obtaining the element.

Silver perrhenate, $AgReO_4$. The silver salt is readily obtained by dissolving silver oxide in hot perrhenic acid, or by adding silver nitrate solution to a hot saturated potassium perrhenate solution. As the silver salt is only one-third as soluble as that of potassium, silver perrhenate is readily separated as a micro-crystalline precipitate.

Silver perrhenate forms small needles which darken on exposure to light. It melts at about 43° C., decomposing at 455° C., giving off a gas and leaving a brown solid (**18**). Its density is 7·05. The salt was prepared and used by Hönigschmid and Sachtleben in determining the atomic weight of rhenium (see p. 25).

Thallium perrhenate, $TlReO_4$. This is the least soluble of the metallic perrhenates (solubility 1·6 g./l. at 20° C.), and it is formed, both as very small needles and as a precipitate, upon adding thallium salt solutions to those of a soluble perrhenate. Jaeger and Beintma (**19**) state that it forms rhombic plates, flattened parallel. It appears green in transmitted light and white in reflected light. Its formation was used by Krauss and Steinfeld for estimating rhenium (**20**). It melts at 527 ± 3° C. with decomposition and shows a transition point (anisotropic change) at about 120–123° C.

Barium perrhenate, $Ba(ReO_4)_2$. This salt is obtained by dissolving barium hydroxide in perrhenic acid, and it crystallizes with two molecules of water of crystallization as small platelets which are fairly soluble in water and which become anhydrous *in vacuo*. The anhydrous salt can be fused without decomposing.

Copper perrhenate, $Cu(ReO_4)_2, 5H_2O$. This light blue salt was first described by Briscoe, Robinson and Rudge (**21**), who obtained it by dissolving copper carbonate in perrhenic acid. The crystals lose one molecule of water of crystallization on standing over calcium chloride. The anhydrous salt, obtained on heating the crystals, is white or slightly greenish, and absorbs water from the air, first forming a hemi-hydrate ($\frac{1}{2}H_2O$) which then passes to the tetrahydrate.

Addition of ammonium hydroxide produces the ammine which separates as deep blue crystals, $Cu(ReO_4)_2, 4NH_3$, only moderately soluble in water and unchanged on heating to 100° C. Stronger heating, however, decomposes them.

Manganese perrhenate, $Mn(ReO_4)_2$, $3H_2O$. This formed deliquescent, pink prisms and was obtained by dissolving manganese carbonate in perrhenic acid. The anhydrous salt decomposed above 300° C., giving off rhenium heptoxide.

Nickel perrhenate, $Ni(ReO_4)_2$, $5H_2O$, was similarly obtained from the carbonate and perrhenic acid as blue-green crystals containing five molecules of water of crystallization, although the water content could vary. Dehydration occurred in stages, the tetrahydrate being formed before the cream-yellow anhydrous salt was left at 170° C.

Treatment of the solution with ammonium hydroxide produced two ammines, namely, the hexammine, $Ni(ReO_4)_2$, $6NH_3$, which separated as lilac crystals, and the tetrammine, $Ni(ReO_4)_2$, $4NH_3$, a pale blue substance remaining in solution. When heated these ammines left nickel oxide.

Cobalt perrhenate, $Co(ReO_4)_2$, $5H_2O$. This salt was similarly obtained from cobalt carbonate and perrhenic acid. The pink solution yielded dark pink crystals that were readily soluble in water and also lost their water of crystallization more easily than the copper or nickel salts. A trihydrate was formed as an intermediate product. The anhydrous cobaltous perrhenate is a purple-blue solid.

When ammonia gas is led into hot cobalt perrhenate solution bright violet crystals of the tetrammine, $Co(ReO_4)_2$, $4NH_3$, are formed. If these are heated they leave a green powder containing rhenium. Other cobalt-ammine perrhenates are described below.

Neodymium perrhenate, $Nd(ReO_4)_3$, has been obtained as rose-coloured needle prisms by dissolving the hydroxide in perrhenic acid.

Lanthanum perrhenate, $La(ReO_4)_3$, is made by dissolving the rare earth oxide in perrhenic acid. The salt crystallizes with two or with three molecules of water of crystallization, the transition point being 30° C. Both salts are prismatic and monoclinic, but the dihydrate has a pinacoidal habit (**22**).

Nitron perrhenate, $C_{20}H_{16}N_4$, $HReO_4$, was first mentioned by Geilmann and Voigt (**23**), who obtained it as a very difficultly soluble precipitate on adding nitron acetate to an acetic acid solution of a perrhenate. The solubility is only 0·018 g. in 100 c.c. of water at 15° C., and this is depressed to almost zero in 3 % nitron acetate solution. A similar precipitate is given by molybdates, but in their absence the formation of nitron perrhenate can be used for the quantitative estimation of rhenium.

Methylene blue perrhenate has been prepared as deep blue needles which show dichroism from yellow to violet-red. Among the other organic salts of perrhenic acid that have been obtained are those of trypaflavine (yellow-red needles showing dichroism from red to green), brucine (colourless needles or prisms) and strychnine.

9-Amino-acridine hydrochloride and 3:6-diamino-acridine hydrochloride also gave crystalline products with perrhenic acid. In contrast with the potassium salt, all the organic perrhenates give a good platinum wire flame coloration test for rhenium.

In addition to the above salts a number of others were obtained and described by Wilke-Dorfürt and Gunzert (**24**). They include

$Ca(ReO_4)_2$	$Sr(ReO_4)_2$	$Zn(ReO_4)_2$
$Ag(NH_3)_2ReO_4$	$Zn(NH_3)_2ReO_4$	$Co''(NH_3)_4ReO_4$
$Co'''(NH_3)_6ReO_4$	$Cd(ReO_4)_2$	$Cr(NH_3)_6ReO_4$
$CrCO(NH_3)_6ReO_4$	$Ag(Py_4)ReO_4$	$Cu(Py_4)ReO_4$

The following more complex salts were also prepared:

Cobalt hexammine perrhenate	$[Co(NH_3)_6](ReO_4)_3$
Cobalt tetrammine perrhenate	$[Co(NH_3)_4](ReO_4)_2$
Chromium hexammine perrhenate	$[Cr(NH_3)_6](ReO_4)_3$
Chromium hexa-urea perrhenate	$[Cr(CO(NH_2)_2)_6](ReO_4)_3$
Zinc tetrammine perrhenate	$[Zn(NH_3)_4](ReO_4)_2$
Cadmium tetrammine perrhenate	$[Cd(NH_3)_4](ReO_4)_2$
Silver diammine perrhenate	$[Ag(NH_3)_2]ReO_4$
Copper tetrapyridine perrhenate	$[Cu(C_5H_5N)_4](ReO)_2$
Silver tetrapyridine perrhenate	$[Ag(C_5H_5N)_4]ReO_4$
Nitrosyl perrhenate	$(NO)ReO_4$

The following cobaltammine perrhenates have been described by Neusser (**25**):

$[Co(NH_3)_6](ReO_4)_3$, $1{\cdot}5H_2O$	Cobalt-hexammine (luteo) perrhenate
$[Co(NH_3)_6(H_2O)](ReO_4)_3$, $3H_2O$	Cobalt-aquo-pentammine (roseo) perrhenate
$[Co(NH_3)_5Cl](ReO_4)_2$	Cobalt-chloro-pentammine (purpureo) perrhenate
$[Co(NH_3)_5NO](ReO_4)_2$	Cobalt-nitro-pentammine (xantho) perrhenate
$[Co(NH_3)_4C_2O_4]ReO_4$	Cobalt-oxalato-tetrammine perrhenate

A short description of the more important of these bodies is now given.

Cobalt tetrammine perrhenate, $[Co(NH_3)_4](ReO_4)_2$, is obtained by leading ammonia gas into a concentrated cobalt perrhenate solution to which a little hydroxylamine hydrochloride has been added, air having been displaced from the apparatus by nitrogen. At first a red precipitate appears but, on shaking, this turns to a purple-red crystalline mass and under the microscope is seen to consist of fine tetrahedral crystals. The precipitate must be filtered in an atmosphere of nitrogen and washed with alcoholic ammonia. It is dried over solid sodium hydroxide.

Cobalt hexammine perrhenate, $[Co(NH_3)_6](ReO_4)_3$, is readily obtained from 0·001 g.mol. of cobalt hexammine chloride dissolved in 5 c.c. of water and excess of perrhenic acid solution of equal concentration. The microcrystalline precipitate was filtered off, drained and washed with a little absolute alcohol and dried. The compound is a yellow, crystalline powder which loses its two molecules of water of crystallization when kept *in vacuo* over phosphorus pentoxide. Its solubility is 0·469 g./l. at room temperatures.

Chromium hexammine perrhenate, $[Cr(NH_3)_6](ReO_4)_3$, was obtained from 0·001 g.mol. of chromium hexammine nitrate in 15 c.c. of water on treating with excess of a perrhenic acid solution. It forms small lemon-yellow prisms containing two molecules of water of crystallization which are lost on standing over phosphorus pentoxide *in vacuo*. Its solubility is 0·684 g./l. at 20° C.

Chromium hexa-urea perrhenate, $[Cr(CO(NH_2)_2)_6](ReO_4)_3$, was obtained from chromium hexa-urea chloride and excess of concentrated perrhenic acid. Crystals separated when the solution was cooled and the deposit was washed with perrhenic acid, redissolved in water and again precipitated with perrhenic acid, washed with the dilute acid and finally with absolute alcohol, and dried over calcium chloride. The salt forms green needle prisms, soluble in both alcohol and water (17·86 g./l. in water at 20° C.).

Zinc tetrammine perrhenate, $[Zn(NH_3)_4](ReO_4)_2$, was obtained when 0·004 g.mol. of perrhenic acid was treated with the calculated amount of zinc carbonate, and the solution of zinc perrhenate was concentrated on a water-bath prior to the addition of concentrated ammonium hydroxide which precipitated the zinc tetrammine perrhenate. This was filtered off, washed in a little concentrated ammonium hydroxide and dried *in vacuo* over solid sodium hydroxide.

Zinc tetrammine perrhenate is a white crystalline powder, soluble in solutions of ammonia.

Cadmium tetrammine perrhenate, $[Cd(NH_3)_4](ReO_4)_2$, was prepared in the same way as the zinc salt and formed colourless regular crystals, also soluble in aqueous ammonia solutions.

Silver diammine perrhenate, $[Ag(NH_3)_2]ReO_4$, was obtained by dissolving the difficultly soluble silver perrhenate in concentrated ammonium hydroxide solution (1 g. in 8 c.c.) and then cooling the solution in a freezing mixture of ice and salt until a mass of fine crystals separated. These were filtered off, washed with ice-cold ammonium hydroxide and dried over solid sodium hydroxide.

The salt forms colourless monoclinic prisms not specially affected by light.

Copper tetrapyridine perrhenate, $[Cu(C_5H_5N)_4](ReO_4)_2$, was obtained from 0·001 g.mol. of cupric chloride by warming with 0·01 g.mol. of pyridine and then treating with 0·002 g.mol. of perrhenic acid. On cooling, a blue salt crystallized out and could be recrystallized from hot dilute pyridine containing perrhenic acid.

Copper tetrapyridine perrhenate forms blue needle prisms which are fairly soluble in water, namely 5·555 g./l. at 20° C.

Silver tetrapyridine perrhenate, $[Ag(C_5H_5N)_4]ReO_4$, separated as an oil from a mixture of aqueous silver nitrate, pyridine and perrhenic acid. Upon cooling, this oil set to a mass of needle crystals. When dried over solid sodium hydroxide the salt lost pyridine even in an atmosphere of pyridine vapour.

Nitrosyl perrhenate, $(NO)ReO_4$, was obtained by leading a current of dry nitric oxide into concentrated (syrupy) perrhenic acid. No reaction occurred if dilute acid was used, or with rhenium heptoxide and dry nitric oxide so that a little water seemed necessary.

Nitrosyl perrhenate forms colourless, very deliquescent crystals, easily decomposed by water into the free acid and nitric oxide.

Among other curious pyridyl derivatives are the 2:2′-dipyridyl perrhenate, $C_5H_4N.NC_5H_4.HReO_4$, and the 2:2′:2″-tripyridyl salt, $C_5H_4N.C_5H_4N.C_5H_4N.HReO_4$. The dipyridyl salt was obtained by Turkiewicz (26) from a solution of the base in dilute acetic acid and perrhenic acid as leaf-like clusters of colourless, shiny crystals. The tripyridyl salt is likewise obtained on mixing an acetic acid solution of the base with perrhenic acid, when it separates as a white precipitate insoluble in water.

The Rhenates. Very little is yet known concerning the rhenates, M_2ReO_4. When rhenium is added to molten caustic alkali a yellow mass is obtained. This probably contains a rhenate, e.g. Na_2ReO_4 or K_2ReO_4, and the alkaline solution obtained when the mass is lixiviated gives a precipitate with silver nitrate solution and with solutions of barium salts. With excess of alkali hydroxide the solutions remain yellow and if evaporated over phosphorus pentoxide yellow crystals separate.

The silver rhenate precipitate is white and dissolves in nitric acid.

When barium perrhenate is heated to 300° C. in a current of nitrogen, black rhenium dioxide is formed. If the product is dissolved in 5 % nitric acid solution and treated with sodium or potassium hydroxides until alkaline, four-sided plates of barium rhenate, $BaReO_4$, separate. These darken on exposure, possibly through hydrolysis.

REFERENCES

(1) Briscoe, Robinson and Stoddart. *J. Chem. Soc.* 1931, p. 666.
(2) Hölemann. *Z. anorg. Chem.* 1934, **220**, 33.
(3) Feit. *Z. angew. Chem.* 1931, **44**, 65.
(4) Geilmann and Weibke. *Z. anorg. Chem.* 1931, **195**, 289.
(5) Hagen and Sieverts. *Z. anorg. Chem.* 1932, **208**, 367.
(6) Jakob and Jeżowska. *Ber. dtsch. chem. Ges.* 1933, **66**, 461.
(7) Noddack, I. and W. *Das Rhenium*, p. 45.
(8) Scharnow. *Z. anorg. Chem.* 1933, **215**, 184.
(9) Noddack, I. and W. *Das Rhenium*, p. 129.
(10) Becker and Moers. *Metallwirtschaft*, 1930, **9**, 1063.
(11) Vorländer and Dalichau. *Ber. dtsch. chem. Ges.* 1933, **66**, 1534.

(**12**) Hölemann and Kleese. *Z. anorg. Chem.* 1938, **237**, 172.
(**13**) Biltz. *Z. anorg. Chem.* 1933, **214**, 225.
(**14**) Puškin and Kovać. *Bull. soc. chim. Yugoslavie*, 1931, **2**, 111.
(**15**) Noddack, I. and W. *Z. angew. Chem.* 1931, **44**, 215.
(**16**) Roth and Becker. *Z. phys. Chem.* 1932, A, **159**, 27.
(**17**) Hölemann. *Z. anorg. Chem.* 1931, **202**, 277.
(**18**) Vorländer, Hollatz and Fischer. *Ber. dtsch. chem. Ges.* 1932, **65**, 535.
(**19**) Jaeger and Beintma. *Proc. K. Akad. Wet. Amst.* 1933, **36**, 523.
(**20**) Krauss and Steinfeld. *Z. anorg. Chem.* 1931, **197**, 52.
(**21**) Briscoe, Robinson and Rudge. *J. Chem. Soc.* 1931, p. 2211.
(**22**) Vezirzade. *Izv. Azerbaijan. Akad. Nauk*, 1940, **1**, 14.
(**23**) Geilmann and Voigt. *Z. anorg. Chem.* 1931, **195**, 298.
(**24**) Wilke-Dorfürt and Gunzert. *Z. anorg. Chem.* 1933, **214**, 369.
(**25**) Neusser. *Z. anorg. Chem.* 1937, **230**, 253.
(**26**) Turkiewicz. *Roczn. Chem.* 1932, **12**, 589.

CHAPTER V

THE HALOGEN COMPOUNDS OF RHENIUM

THE following halides of rhenium have been reported at different times since 1925:

	$ReCl_3$	$ReBr_3$	
ReF_4	?$ReCl_4$	$ReBr_4$	ReI_4
ReF_6	$ReCl_5$		
	?$ReCl_6$		
	?$ReCl_7$		

The following oxyhalides have been announced:

ReO_3F	ReO_3Cl	ReO_3Br
ReO_2F_2	?ReO_2Cl_3	ReO_2Br_2
$ReOF_4$	?$ReOCl_2$	$ReOBr_2$
	$ReOCl_4$	

THE FLUORIDES OF RHENIUM

Rhenium hexafluoride, ReF_6. This compound was prepared by Ruff, Kwasnik and Ascher (1) by the action of fluorine on rhenium at 125° C. At − 180° C. it turned to bright yellow crystals; above 0° C. they were lemon-yellow in colour and melted at 18·8° C. to a yellow-brown liquid which boiled at 47·6° C. and had a density of 3·61. The vapour was almost colourless.

Rhenium hexafluoride is slowly volatilized in dry air and has a vapour density of 300. Its heat of formation is 275 kg.cal. Analysis shows that it contains 62·12 % of rhenium and its formation is independent of pressure and of the amount of fluorine used. It is reduced to rhenium tetrafluoride by hydrogen at 200° C., by carbon monoxide at 300° C., by sulphur dioxide at 400° C. and by rhenium itself at 400–500° C. Metals like copper, gold, zinc and aluminium also reduce the hexafluoride to the tetrafluoride.

Oxygen, peroxides and potassium permanganate convert it into the oxytetrafluoride and trioxyfluoride, ReO_3F. With water rhenium hexafluoride gives hydrated rhenium dioxide and perrhenic and hydrofluoric acids,

$$3ReF_6 + 10H_2O = ReO_2 + 2HReO_4 + 18HF.$$

It reacts with grease, glass (very readily with silica), petroleum ether and with other organic compounds; with silica the reaction is

$$SiO_2 + 2ReF_6 = 2ReOF_4 + SiF_4.$$

Rhenium tetrafluoride, ReF_4. Rhenium tetrafluoride is a solid, melting at 124·5° C., obtained by reduction of the hexafluoride as described above.

The Chlorides of Rhenium

Rhenium heptachloride, $ReCl_7$. In early work on the isolation of compounds of rhenium frequent reference was made to a volatile green higher chloride obtained when crude rhenium preparations were acted on by concentrated hydrochloric acid (2). This was assumed to be the heptachloride, although I. and W. Noddack in 1931 (3) regarded it as the hexachloride. It may equally well have been an oxychloride, e.g. ReO_3Cl, analogous with the corresponding manganese oxychloride, MnO_3Cl.

A product also supposed to be the heptachloride was obtained by passing chlorine over rhenium, although this unstable volatile green substance was not the main product of the reaction. When absorbed in water hydrochloric and perrhenic acids were undoubtedly formed.

Attempts to absorb the alleged heptachloride in ether and other organic solvents were not successful, and it is still doubtful whether the heptachloride of rhenium has ever been isolated.

Rhenium hexachloride, $ReCl_6$. Early references to a hexachloride of rhenium were also frequent, but, as with the heptachloride, its existence is still doubtful. Yost and Shull (4) stated that it was obtained from rhenium when the element was heated in excess of chlorine at about 650° C. They stated that it could not be obtained free from the tetrachloride, but as the existence of this too is now considered doubtful, whilst a trichloride (which they did not observe) is produced and a pentachloride is another major product of the interaction of the two elements, their work does not appear to be final.

Rhenium pentachloride, $ReCl_5$. This halide is formed when chlorine in excess acts upon rhenium, and it is readily purified by vacuum sublimation (5). When its vapour is heated it dissociates into the trichloride and chlorine, resembling phosphorus pentachloride in this respect. Sublimation is not possible at atmospheric pressure without decomposition.

Rhenium pentachloride is a deep brown-black solid, the vapour being dark red-brown. When heated with oxygen it gives chlorine and some oxychlorides,

$$16ReCl_5 + 14O_2 = 10ReOCl_4 + 6ReO_3Cl + 17Cl_2.$$

On heating with potassium chloride the rhenichloride may be formed with evolution of chlorine,

$$2ReCl_5 + 4KCl = 2K_2ReCl_6 + Cl_2.$$

With water and with caustic alkalis a complex decomposition occurs. At first an acid, H_2ReOCl_5, appears to be formed; with alkalis the final reaction may be expressed,

$$3ReCl_5 + 16NaOH = 2Re(OH)_4 + NaReO_4 + 15NaCl + 4H_2O.$$

Alkaline chlororhenate (rhenichloride) is also probably formed at the same time. In estimating the percentage of rhenium in this or other halides by collecting the hydrated rhenium dioxide (which is the same as $Re(OH)_4$) produced, it is found that if carbon dioxide is subsequently bubbled through the solution the precipitate filters much more easily. According to Klemm and Frischmuth (6) ammonia also reacts with rhenium pentachloride (see below).

? *Rhenium tetrachloride*, $ReCl_4$. It has been stated (7) that the interaction of rhenium and chloride produced the tetrachloride, and at one time it was supposed that this was the main product. Probably this view of the reaction was due to the small amount of reactants used and to the difficulties encountered in preparing a pure single product. The reaction begins at ordinary temperatures and becomes rapid as the temperature approaches 600° C. The dark volatile product was assumed to be the tetrachloride and a similar substance (or mixture) was obtained by heating rhenium dioxide with sugar charcoal in a stream of chlorine, as an almost black sublimate. I. and W. Noddack (*Das Rhenium*, p. 65) also describe the tetrachloride in some detail and assign to it a melting-point of 180° C. and a boiling-point of about 250° C. Yost and Shull gave its molecular weight (from vapour-density determinations) as 472, and this would suggest that it was partly polymeric (Re_2Cl_8), since the molecular weight of $ReCl_4$ should be 328·14.

The very dark brown solid appeared to be stable in dry air and sublimed on heating. Mixed with a little water it gave a blue liquid which gradually hydrolysed, especially with more water, giving the dioxide and hydrochloric acid among other products.

Geilmann, Wrigge and Biltz (5) failed to prepare the tetrachloride, whereas they readily obtained the trichloride and pentachloride. Analysis that had shown a Re : Cl ratio of 1 : 4 can be accounted for by admixture of some $ReOCl_4$ with the pentachloride and trichloride.

When treated with hydrochloric acid the supposed tetrachloride gave a deep green solution. This probably contained chlororhenic acid, H_2ReCl_6.

Rhenium trichloride, $ReCl_3$. This halide arises from the initial action of chlorine on rhenium, and it is also a product of the thermal decomposition (in an atmosphere of nitrogen) of the pentachloride which is generally the main product of the action. It results sometimes in other reactions, e.g. in the action of heat on silver rhenichloride *in vacuo* at 350–400° C.,

$$2Ag_2ReCl_6 = 4AgCl + ReCl_3 + ReCl_5.$$

Rhenium trichloride is a reddish black or violet, many-faced hexagonal, lustrous, crystalline solid, purified by sublimation (its vapour is dark green) at 500–550° C. It gives intensely red solutions in water, acetone, ether, glacial acetic acid and in aqueous hydrochloric acid, with which it forms $HReCl_4$. It is incompletely precipitated by hydrogen sulphide and, at first, its aqueous solution gives no precipitate with silver nitrate; addition of mineral acids, however, soon brings about hydrolysis.

Ammonium hydroxide first gives (6) a violet-red precipitate, but this dissolves in excess to a blue solution. The compounds $ReCl_3,14NH_3$; $ReCl_3,7NH_3$ and $ReCl_3,6NH_3$ are reported to be formed. Thiocyanates give a yellow colour and ferrocyanides a blue one with the trichloride.

Rhenium trichloride is reduced by hydrogen to rhenium and hydrochloric acid gas at 250–300° C., no lower chloride being formed in the process. When heated in air or oxygen it gives oxychlorides and chlorine,

$$6ReCl_3 + 7O_2 = 4ReO_3Cl + 2ReOCl_4 + 3Cl_2.$$

Double salts were obtained by Biltz (8) and others on adding rubidium or caesium chlorides and with pyridine hydrochloride. On heating with potassium chloride *in vacuo* at 600° C. it gave potassium rhenichloride and rhenium,

$$6KCl + 4ReCl_3 = 3K_2ReCl_6 + Re.$$

A compound of rhenium trichloride with two molecules of water of crystallization, $ReCl_3,2H_2O$, was obtained in the form of short rods by Wrigge and Biltz (9).

Rhenium tribromide, $ReBr_3$. Hagen and Sieverts (10) found that rhenium combines with bromine vapour at 300° C. and gives a stable green-black sublimate. The vapour is dark green. The tribromide can also be made by dissolving rhenium sesquioxide in hydrobromic acid. I. and W. Noddack found (11) that the tribromide unites with pyridine hydrobromide to form the dark red double compound, $C_5H_5N,ReBr_3,HBr$.

In the presence of oxygen at 400° C. the tribromide yields a blue distillate, probably an oxybromide.

Rhenium tetra-iodide, ReI_4, is reported to be formed when the element is exposed to iodine vapour, but no details have been published (I. and W. Noddack, *Das Rhenium*, p. 67).

THE OXYHALIDES OF RHENIUM

Rhenium trioxychloride, ReO_3Cl. Brukl and Ziegler (12) state that rhenium trioxychloride resulted when the tetrachloride and heptoxide were heated together. Geilmann and Wrigge (13) obtained it in the form of white needles (simultaneously with the oxytetrachloride) from rhenium trichloride and oxygen, or air, on heating. The crude product was slightly blue, or green, but when purified the compound is a colourless liquid, freezing at 3–4° C.

and boiling at 131° C. It can be distilled undecomposed in chlorine or in oxygen. It fumes and hydrolyses in moist air, finally leaving perrhenic acid; the hydrochloric acid also formed volatilizes away.

Rhenium dioxytrichloride, ReO_2Cl_3. This was described by Briscoe, Robinson and Rudge (**14**) as consisting of dark red-brown needles, darkening still more on keeping and melting at 23·9° C. It was prepared (*a*) by heating rhenium in air and chlorine, (*b*) by heating rhenium tetrachloride in oxygen and (*c*) by heating rhenium pentoxide in chlorine. The first method was the most satisfactory.

At 35° C. the liquid had a density of 3·359 and vaporized a little below 300° C. *in vacuo*. At 400° C. its vapour density corresponded with a molecular weight of 332. It could be heated with rhenium to 260° C. without further action. A complex decomposition took place with water and it fumed in moist air.

Rhenium oxytetrachloride, $ReOCl_4$. This is a second product of Brukl and Ziegler's reaction (**12**) between the tetrachloride (? trichloride) and heptoxide, when the halide is in excess. It can also be made, according to these authors, by heating the halide in a stream of air, or oxygen, to 130° C. By heating the trichloride in oxygen or the pentachloride in air, rhenium oxytetrachloride is most conveniently obtained,

$$6ReCl_3 + 5O_2 = 4ReOCl_4 + 2ReO_3Cl_2.$$

Rhenium oxytetrachloride forms a brown, crystalline mass, m.p. 29–30·5° C., b.p. 223° C. (with slight decomposition). With water it decomposed into hydrochloric and perrhenic acids and hydrated rhenium dioxide. With cold concentrated hydrochloric acid it gives a brown solution which may contain an unstable acid, H_2ReOCl_6, since Brukl and Ziegler claim to have obtained a salt, K_2ReOCl_6, from it. The solution eventually contains chlororhenic acid together with perrhenic and hydrochloric acids,

$$3H_2ReOCl_6 + 5H_2O = H_2ReCl_6 + 2HReO_4 + 12HCl.$$

Careful treatment of the oxytetrachloride with ammonia gave the compound $ReO(NH_2)_2Cl_2$. This solid was stable in dry air but decomposed when heated above 400° C., leaving rhenium dioxide and free rhenium. It is hydrolysed even at 0° C. to $ReO(OH)_2(NH_2)_2$, which lost water *in vacuo* at 100° C., yielding $ReO_2(NH_2)_2$.

Rhenium oxydichloride, $ReOCl_2$ (?). Brukl and Plettinger (15) found that when water was carefully added to rhenium oxytetrachloride in organic solvents such as carbon tetrachloride, chloroform, acetone, ether or benzene, a blue precipitate was formed. The rhenium and chlorine were in the ratio 1 : 2, but there were varying amounts of oxygen and hydroxyl. The product showed great susceptibility to heat, and this made it impossible to separate the components of the reaction product which was supposed to contain rhenium oxydichloride.

No oxyiodide has been described, but Ruff, Kwasnik and Ascher (1) mention that *rhenium oxytetrafluoride*, $ReOF_4$, resulted from the hydrolysis of the hexafluoride. They later (16) isolated it as colourless crystals, m.p. 39·7° C., giving a colourless liquid of density 3·72 and boiling at 62·7° C.

Rhenium dioxydifluoride, ReO_2F_2, is described as a colourless solid, m.p. 156° C., and Ruff and Kwasnik state that there are indications of the existence of an oxydifluoride and a trioxyfluoride as well.

The oxybromides ReO_3Br and ReO_2Br_2 were made by Brukl and Ziegler (17) by the interaction of bromine and rhenium heptoxide.

Rhenium trioxybromide is produced as a blue-black solid in greater quantity than the *dioxydibromide*. The former melts at 39·5° C. and boils at 163° C. The dioxydibromide retained 1–2 % of bromine that could not be removed since the compound decomposes at 60–70° C., which is just before it melts. The remaining oxybromide of rhenium, $ReOBr_2$, was mentioned by Feit (18) as forming a red solution.

The Double Chlorides of Rhenium

Early attempts to prepare rhenichlorides, the salts of chlororhenic (hexa-chlororhenic) acid, H_2ReCl_6, were not very successful, possibly because the alleged rhenium tetrachloride used in the experiments was actually impure trichloride. Moreover,

Briscoe and his collaborators (19) found that potassium rhenichloride was not stable in solution.

Besides salts of the type M_2ReCl_6, others with the formula $MReCl_4$ have been obtained.

Potassium rhenichloride, K_2ReCl_6, can be prepared by the action of hydrochloric acid on a mixture of potassium iodide and perrhenate (20),

$$KReO_4 + 3KI + 8HCl = K_2ReCl_6 + 2KCl + 3I + 4H_2O.$$

It is not easily obtained from rhenium chlorides and potassium chloride, although it can be made by heating the pentachloride with potassium chloride, some chlorine being evolved in the process,

$$2ReCl_5 + 4KCl = 2K_2ReCl_6 + Cl_2.$$

Briscoe, Robinson and Stoddart also obtained it by fusing rhenium with potassium chloride in a stream of chlorine.

Potassium rhenichloride forms yellow-green, octahedral microcrystals (sometimes larger ones can be obtained) of density 3·34; it is not very soluble in water but dissolves fairly readily in hydrochloric acid. The aqueous solution ionizes into potassium and rhenichloride ions, but the solution soon hydrolyses. According to Briscoe and others the solution is unstable above 25° C. The salt is unaffected by sulphurous or hypophosphorous acids but is immediately reduced by zinc and hydrochloric acid. Ammonium hydroxide precipitated hydrated rhenium dioxide.

Manchot and Düsing showed (21) that electrolytic reduction of its dilute solutions yields complexes of tervalent rhenium. The reduction products rapidly take up oxygen again from potassium permanganate or hydrogen peroxide.

The complex salts, K_3ReCl_6 and $K_4Re_2Cl_{11}$, were obtained by Krauss and Steinfeld (22), as well as the commoner K_2ReCl_6, by the action of hydrochloric acid on a mixture of potassium perrhenate and iodide.

The following is an account of some oxy-salts.

Potassium oxyrhenichloride, K_2ReOCl_5, has been described by Jakob and Jeżowska (23) as resulting from the same interaction in concentrated hydrochloric acid. Crystals having a composition agreeing with the above formula separated with the iodine simultaneously liberated from the iodide and perrhenate. After filtering off the crystals on to glass-wool they were freed from iodine by washing with ether and were then dried on a porous plate in an exsiccator.

This potassium oxyrhenichloride, or pentachlororhenate, was free from both perrhenate and ordinary rhenichloride. It was in the form of yellow-green crystals which remained unchanged in dry air, but they turned black in moist air, owing to the formation of hydrated rhenium dioxide as a result of hydrolysis. The crystals were only slightly soluble in concentrated hydrochloric acid, in which they gave a yellow solution. The compound was more soluble in dilute acid, giving a green solution, but further dilution caused the colour to darken still more to bluish green and finally black (with separation of rhenium dioxide).

When this oxyrhenichloride was heated with concentrated hydrochloric acid it was converted into the ordinary rhenichloride with some perrhenate. In concentrated sulphuric acid it yielded a red-brown solution which turned violet on warming but eventually became colourless. With hydrogen sulphide rhenium sulphide was obtained.

Ammonium oxyrhenichloride, $(NH_4)_2ReOCl_5$, was obtained in a similar way from ammonium perrhenate and ammonium iodide in concentrated hydrochloric acid, in the form of greenish yellow crystals, darker than the potassium salt with which they are isomorphous. Other oxy- and hydroxychlorides have been reported. Thus, Krauss and Steinfeld mention $K_4Re_2OCl_{10}$ and $Ag_4Re_2OCl_{10}$; Jakob and Jeżowska prepared $K_2Re(OH)_2Cl_5$ by electrolysis of a hydrochloric acid solution of potassium perrhenate and iodide. The corresponding caesium and rubidium salts, $Cs_2Re(OH)_2Cl_5$—light yellow micro-crystals—and $Rb_2Re(OH)_2Cl_5$—green crystals—were described in 1932 by Krauss and Dahlmann (24). These substances, if they are pure compounds, may be regarded as products of the partial hydrolysis of the rhenichlorides.

In the course of an examination of the hydrolysis of potassium rhenichloride, Geilmann and Hurd (25) devised a means for estimating the rhenichloride in the presence of perrhenate. The rhenichloride is quantitatively precipitated by the organic reagent, tetron (*N*.*N*′-tetramethyl-*o*-toluidine), whereas perrhenates are not affected and may afterwards be precipitated with nitron, or with hydrogen sulphide.

Rubidium and *caesium rhenichlorides*, Rb_2ReCl_6 and Cs_2ReCl_6, were obtained as green, micro-crystalline precipitates, sparingly soluble in water, by adding the alkali chlorides to chlororhenic acid or to a hydrochloric acid solution of the potassium salt. The same salts resulted when the double compounds of alkali halide and rhenium trichloride were heated, e.g.

$$6RbReCl_4 = 3Rb_2ReCl_6 + Re + 2ReCl_3.$$

Rubidium tetrachlororhenate (or rubidium rhenium chloride), $RbReCl_4$, was itself obtained as a fine, deep red, crystalline powder, easily soluble in water and dilute acids. It showed the general reactions of rhenium trichloride.

The *caesium* compound, $CsReCl_4$, formed minute bi-pyramids, less soluble in water than the rubidium salt. Both these bodies resulted from the union of the alkali chloride and rhenium trichloride (**26**).

Silver rhenichloride, Ag_2ReCl_6. This compound was first described by Briscoe, Robinson and Stoddart (**19**), who obtained it as an orange precipitate when silver nitrate solution was added to potassium rhenichloride solutions. Geilmann and Wrigge (**26**) found that this salt decomposed on heating at 350–400° C. *in vacuo*, leaving silver chloride and evolving rhenium trichloride and pentachloride.

Yellow precipitates of thallium rhenichloride (hexachlororhenate), Tl_2ReCl_6, and mercurous rhenichloride, Hg_2ReCl_6, are known, whilst Krauss and Steinfeld (**22**) also claimed to have prepared salts with the formulae Tl_3ReCl_6 and Ag_3ReCl_6. Toluidine, quinoline and pyridine salts of chlororhenic acid have been described (**27**). They are yellow-brown crystalline solids which decompose below their melting-points and have the general formula, $B_2H_2ReCl_6$.

Turkiewicz (**28**) obtained 2:2′-dipyridyl rhenichloride (hexachlororhenate), $C_5H_4N.C_5H_4N,H_2ReCl_6$, as yellow needles by mixing the base and chlororhenic acid in concentrated hydrochloric acid.

A *bis*-2:2′-dipyridyl rhenichloride, $(C_5H_4N.C_5H_4N)_2H_2ReCl_6$, and a tripyridyl salt, $C_5H_4N.C_5H_4N.C_5H_4N.H_2ReCl_6$, have also been described. The bis-salt was obtained as pale green, acicular needles from a dilute acetic acid solution of the base and potassium rhenichloride. The crystals are only sparingly soluble in water.

The tripyridyl compound was also only very slightly soluble in water and was obtained in a similar way.

A number of complex oxychlorides of quadrivalent rhenium were prepared by Jeżowska and Iodka (**29**) from perrhenic acid in 30 % hydrochloric acid with the metal iodide. The mixtures were kept at 20° C. for 24 hr. in an atmosphere of carbon dioxide. The salts obtained included those with the following formulae:

$$K_4Re_2OCl_{10}, \quad (NH_4)_4Re_2OCl_{10}$$

and the quinoline salt $(C_9H_7N)_4Re_2OCl_{10}$;

and $K_2Re(OH)Cl_5$, $Rb_2Re(OH)Cl_5$,

$$(NH_4)_2Re(OH)Cl_5 \quad \text{and} \quad (C_9H_7N)_2Re(OH)Cl_5.$$

The first products are the hydroxychlorides, then the oxychlorides, and finally the solution changes to one containing rhenichlorides (hexachlororhenates), M_2ReCl_6.

Bromorhenic acid, H_2ReBr_6. When rhenium dioxide was treated with hydrobromic acid of 48 % strength (**30**) a deep yellow solution resulted. Addition of potassium bromide caused the separation of violet-red crystals of the rhenibromide (hexabromorhenate), K_2ReBr_6; addition of rubidium bromide similarly precipitated the salt Rb_2ReBr_6, as dark red micro-crystals. The caesium salt, Cs_2ReBr_6, was believed to have formed in the same way.

A compound, $Cs_2Re(OH)Br_5$, was reported by Krauss and Dahlmann (**24**).

Potassium rheni-iodide (*Hexa-iodorhenate*), K_2ReI_6. This compound was first obtained by Briscoe, Robinson and Rudge (**31**) by the reduction of potassium perrhenate with hydriodic acid in the presence of excess of potassium iodide as brown-black crystals, soluble in water or acetone, giving a deep red, easily hydrolysed solution.

The corresponding sodium and caesium rheni-iodides were prepared by Krauss and Dahlmann as black, shining crystalline solids.

Hexa-iodorhenic acid, H_2ReI_6, has itself not been isolated, but the penta-iodorhenic acid, H_2ReI_5, has been extracted (32) from the products of the interaction of potassium rheni-iodide with fairly strong (20 %) sulphuric acid. The reaction is as follows:

$$K_2ReI_6 + H_2SO_4 = K_2SO_4 + H_2ReI_5 + I.$$

Attempts have been made to form ammine and other complexes with double rhenium halides (33). The salts K_2ReCl_6 and K_3ReCl_6 partly dissolved in anhydrous ammonia to give orange and yellow-green solutions respectively, but no complexes were actually isolated.

Rhenium halides with excess of ethylene diamine did give yellow prisms of $[ReO_2en_2]Cl$. With potassium iodide this gave the corresponding yellow crystalline iodine compound $[ReO_2en_2]I$. The ammine complex formed a sparingly soluble chlorate, a picrate, platinichloride and cobaltinitrite.

From the red hydrochloric acid solution of the ethylene diamine complex, alcohol precipitated the compound $[ReO(OH)en]Cl_2$, from which, in solution, the platinichloride, $[ReO(OH)en_2]PtCl_6$, was obtained with chloroplatinic acid. Sodium cobaltinitrite, when added to the hydrochloric acid solution of the ethylene diamine complex, gave the cobaltinitrite, $[ReO(OH)en_2]Co(NO_2)_6$, and with potassium iodide the product was $[ReO(OH)en_2]I_2$. All these salts were only slightly soluble in water.

Another compound, $[Re(OH)_2en_2]Cl_3$, was isolated in the form of pale blue needle crystals.

The Carbonyl Halides of Rhenium

A new class of rhenium carbonyl halide has been prepared during recent years. Thus Schulten obtained (34) rhenium pentacarbonyl chloride, $Re(CO)_5Cl$, by heating rhenium pentachloride in carbon monoxide under a pressure of 200 atm. in the presence of copper,

$$ReCl_5 + 4Cu + 9CO = Re(CO)_5Cl + 4CuCl.CO.$$

Rhenium pentacarbonyl chloride was extracted with ether and remained as a fine grey crystalline powder, stable in air, insoluble in water but soluble in organic solvents. Instead of the pentachloride it can be made from potassium rhenichloride, which is heated in an autoclave at 200 atm. pressure of carbon monoxide for 30 hr.

Rhenium pentacarbonyl bromide, $Re(CO)_5Br$, was similarly obtained from the rhenibromide, and the rheni-iodide yields the pentacarbonyl iodide, $Re(CO)_5I$.

A variation of the method of preparation has recently been made (35) whereby the element (rhenium) is heated with a dissociable halide in an atmosphere of carbon monoxide under pressure (200 atm.) at 250° C.,

$$Re + CuCl_2 + 6CO = Re(CO)_5Cl + Cu(CO)Cl,$$

$$Re + CuBr_2 + 6CO = Re(CO)_5Br + Cu(CO)Br.$$

They diminish in volatility from the iodo- to the chloro-compound and this is the order of increasing stability. All are quite stable compounds and can be sublimed in carbon monoxide; the halogen is not removed by silver.

Rhenium carbonyl is not formed when the element is heated with carbon monoxide under pressure, but it is obtained in good yield from carbon monoxide and rhenium heptoxide (and even from perrhenates) at 250° C. It forms colourless crystals that can be sublimed and are soluble in organic solvents. It decomposes at 400° C. There is evidence that it is dimeric, i.e. $[Re(CO)_5]_2$.

At high temperatures (240° C.) it reacts with certain organic bases to give derivatives that can be regarded as compounds in which the halogen has been replaced by the base, e.g.

$$[Re(CO)_5]_2 + 4C_5H_5N = 2Re(CO)_3C_5H_5N + 4CO.$$

o-Phenanthroline reacts more readily than pyridine to give an analogous compound.

With rhenium carbonyl halides such compounds as $ReCl(CO)_3$, $2C_5H_5N$ are produced.

REFERENCES

(1) Ruff, Kwasnik and Ascher. *Z. anorg. Chem.* 1932, **209**, 113.
(2) Loring and Druce. *Chem. News*, 1925, **131**, 273.
(3) Noddack, I. and W. *Z. angew. Chem.* 1931, **44**, 215.
(4) Yost and Shull. *J. Amer. Chem. Soc.* 1932, **54**, 4657.
(5) Geilmann, Wrigge and Biltz. *Z. anorg. Chem.* 1933, **214**, 244.
(6) Klemm and Frischmuth. *Z. anorg. Chem.* 1937, **230**, 209.
(7) See *Chem. Weekbl.* 1926, **29**, 57.
(8) Biltz *et al.* *Ann. Phys., Lpz.*, 1934, **511**, 301.
(9) Wrigge and Biltz. *Z. anorg. Chem.* 1936, **228**, 372.
(10) Hagen and Sieverts. *Z. anorg. Chem.* 1933, **215**, 111.
(11) Noddack, I. and W. *Z. anorg. Chem.* 1933, **215**, 129.
(12) Brukl and Ziegler. *Ber. dtsch. chem. Ges.* 1932, **65**, 916.
(13) Geilmann and Wrigge. *Z. anorg. Chem.* 1933, **214**, 288.
(14) Briscoe, Robinson and Rudge. *J. Chem. Soc.* 1932, p. 1104.
(15) Brukl and Plettinger. *Ber. dtsch. chem. Ges.* 1933, **66**, 971.
(16) Ruff and Kwasnik. *Z. anorg. Chem.* 1934, **219**, 65.
(17) Brukl and Ziegler. *Mh. Chem.* 1933, **63**, 329.
(18) Feit. *Angew. Chem.* 1933, **46**, 216.
(19) Briscoe, Robinson and Stoddart. *J. Chem. Soc.* 1931, p. 2263.
(20) Enk. *Ber. dtsch. chem. Ges.* 1931, **64**, 791.
(21) Manchot and Düsing. *Z. anorg. Chem.* 1933, **212**, 21.
(22) Krauss and Steinfeld. *Ber. dtsch. chem. Ges.* 1931, **64**, 2552.
(23) Jakob and Jeżowska. *Z. anorg. Chem.* 1934, **220**, 16 and 337.
(24) Krauss and Dahlmann. *Ber. dtsch. chem. Ges.* 1932, **65**, 877.
(25) Geilmann and Hurd. *Z. anorg. Chem.* 1933, **213**, 336.
(26) Geilmann and Wrigge. *Z. anorg. Chem.* 1933, **214**, 248.
(27) Schmid. *Z. anorg. Chem.* 1933, **212**, 187.
(28) Turkiewicz. *Roczn. Chem.* 1932, **12**, 589.
(29) Jeżowska and Iodka. *Roczn. Chem.* 1939, **19**, 187.
(30) Druce. *Chem. News*, 1923, **126**, 1.
(31) Briscoe, Robinson and Rudge. *J. Chem. Soc.* 1931, p. 3218.
(32) Biltz *et al.* *Z. anorg. Chem.* 1937, **234**, 142.
(33) Lebedinsky and Ivanov-Emin. *J. Gen. Chem.*, Russia, 1943, **13**, 253.
(34) Schulten. *Z. anorg. Chem.* 1939, **243**, 164.
(35) Hieber, Schuh and Fuchs. *Z. anorg. Chem.* 1941, **248**, 243.

CHAPTER VI

THE SULPHIDES, SELENIDES AND THIO-SALTS OF RHENIUM

The action of hydrogen sulphide upon solutions of perrhenic acid and perrhenates varies with the conditions, and at first different investigators appeared to obtain almost contradictory results. Thus, when the gas is passed through neutral potassium perrhenate solutions a yellow solution is obtained. Feit (1) attributed this to the formation of a thio-perrhenate,

$$KReO_4 + H_2S = KReO_3S + H_2O,$$

and he succeeded in isolating a number of unstable thio-salts such as the thallium compounds, $TlReO_3S$ and $TlReS_4$.

These thio-salt solutions are unstable and eventually deposit rhenium sulphides. All the rhenium is, however, not precipitated by hydrogen sulphide unless the solution is distinctly acid.

Rhenium forms two well-defined sulphides, the heptasulphide and the disulphide, Re_2S_7 and ReS_2.

Rhenium heptasulphide, Re_2S_7. This is obtained by the continued action of hydrogen sulphide upon perrhenate solutions

$$2HReO_4 + 7H_2S = Re_2S_7 + 7H_2O.$$

Precipitation is not always quantitative, and Briscoe, Robinson, and Stoddart (2) devised a more satisfactory method for preparing the heptasulphide in a pure state, namely, by the interaction of potassium perrhenate with sodium thiosulphate. No action is visible after boiling a mixture of the solutions of these salts, but when the solution is acidified, rhenium heptasulphide is precipitated along with any sulphur due to excess of thiosulphate. The free sulphur is afterwards removed by means of boiling toluene, the product then being kept for some time *in vacuo* over phosphorus pentoxide.

Rhenium heptasulphide is a dark brown, almost black solid. It has a density of 4·866, and is almost insoluble in water and in alkali sulphides, but it is readily attacked by nitric acid with the ultimate formation of perrhenic acid and sulphur or sulphur dioxide, or even sulphuric acid. The heptasulphide adsorbs certain organic solvents, including benzene and toluene, with swelling.

When heated in absence of oxygen, e.g. in nitrogen or carbon dioxide, sulphur sublimes off, leaving the disulphide

$$Re_2S_7 = 2ReS_2 + 3S.$$

When heated in a stream of hydrogen not only is there a sublimation of some sulphur but at red heat all the sulphur is removed, leaving free rhenium. When the heptasulphide is heated in air it readily ignites, forming a white fume of the heptoxide whilst the sulphur is converted into its dioxide,

$$2Re_2S_7 + 21O_2 = 2Re_2O_7 + 14SO_2.$$

Rhenium disulphide, ReS_2. The elements rhenium and sulphur combine at moderate temperature to form the disulphide. The compound also results, as mentioned above, when the heptasulphide is decomposed in an inert gas. At elevated temperatures the disulphide has been decomposed back into its elements by Juza and Biltz (3), who showed that at 1100–1200° C. the decomposition absorbed 70 kg.cal. The heat of formation is 42 kg.cal. Rhenium disulphide is a black solid of density 7·5. It is stable in air unless heated, when it burns to the oxides of rhenium and sulphur. It is not easily oxidized, but hot nitric acid, even dilute, attacks it. When strongly heated in a stream of hydrogen it is reduced to the metal. The precipitation of rhenium as sulphide from various compounds was studied by Geilmann and others (4).

From perrhenate solutions containing not less than 1 % of hydrochloric or sulphuric acid rhenium heptasulphide can be precipitated by hydrogen sulphide under pressure. The method consists in enclosing the solution of not less than 0·1 g. of rhenium in a pressure flask and saturating it with hydrogen sulphide for 20 min. at room temperature. The closed flask is next heated on a water-bath for a further period of 30 min., cooled, opened and the contents filtered on to asbestos. The residue is rhenium heptasulphide, and the precipitation is so complete that it may be used for estimating the element. From this heptasulphide it is also easy to obtain the disulphide by heating in absence of air to drive off the excess of sulphur.

The rhenium in rhenium trichloride is also completely precipitated by hydrogen sulphide in 1–2 hr. in dilute hydrochloric (cold) or sulphuric (hot or cold) acids, or by pressure precipitation. It is also completely precipitated by ammonium sulphide—the mixture being allowed to stand for an hour or two before being acidified to throw down rhenium disulphide, with some heptasulphide and free sulphur. Potassium rhenichloride solutions are only completely precipitated under pressure with hydrogen sulphide, or by treatment with ammonium or sodium sulphide solutions and then acidifying.

A particularly hard sulphide of rhenium was obtained by adding sulphur to the rhenate obtained by fusing rhenium, or the dioxide, with sodium and potassium carbonates. This dark sulphide was extremely difficult to dissolve in acids.

A compound, ReS_3, was mentioned by the Noddacks (5) in 1931. It was said to lose sulphur at 800° C., leaving the disulphide. No one has apparently succeeded in preparing it since.

Rhenium heptaselenide, Re_2Se_7. Briscoe, Robinson and Stoddart (2) obtained the rhenium selenides analogous to the sulphides. The heptaselenide was prepared by passing hydrogen selenide into potassium perrhenate solution containing potash to absorb any selenium liberated. Rhenium heptaselenide was obtained as a fine black powder which decomposed into diselenide and selenium when heated. It absorbed alcohol and other organic liquids with swelling. Hydrogen reduced it to rhenium.

Rhenium diselenide, $ReSe_2$. This compound was obtained by heating the heptaselenide *in vacuo* at 325–330° C. for 9 hr., selenium being volatilized,

$$Re_2Se_7 = 2ReSe_2 + 3Se.$$

Rhenium diselenide was unattacked except by strong oxidizing acids.

Potassium mono-thio-perrhenate, $KReO_3S$. This compound was obtained by Feit (1) by passing hydrogen sulphide into a saturated aqueous potassium perrhenate solution until no more gas was absorbed. The solution, which contained thio-salts, decomposed on keeping, but it was possible to isolate the very soluble mono-thio-salt (soluble also in alcohol) as a yellowish solid, by evaporation. The formula is based upon determinations of the sulphur and potassium in the compound.

It reacted with bromine water, reforming the perrhenate. When its solutions were added to those of salts of lead, copper, mercury and thallium, dark precipitates were obtained, some of them being stable. The lead salt, $Pb(ReO_3S)_2$, appeared as a red precipitate, but eventually decomposed leaving lead sulphide. The copper salt also easily decomposed. With mercuric chloride a white precipitate preceded the appearance of yellow mercury thio-perrhenate, $Hg(ReO_3S)_2$.

Silver nitrate at once gave the sulphide

$$KReO_3S + 2AgNO_3 + H_2O = KReO_4 + Ag_2S + 2HNO_3.$$

Thallium mono-thio-perrhenate, $TlReO_3S$, formed heavy golden yellow needles with a solubility of 1·02 g./l. These may be crystallized from hot water.

More or less pure ammonium, rubidium and caesium thio-perrhenates have been obtained, and Feit prepared the unstable mono-thio-perrhenic acid, $HReO_3S$, which, however, soon deposited rhenium sulphide, leaving some perrhenic acid behind in solution.

Rhenium-tungsten alloys. From a study of the fusion diagrams of mixtures of rhenium and tungsten, Becker and Moers (6) concluded that compounds represented by the formulae WRe, WRe_2 and W_2Re_3 most probably exist. There was one eutectic at 2892° C. and another at 2822° C.

Rhenium also reacts with phosphorus at 750° C. Tensimetric and X-ray examinations (7) point to the existence of compounds possessing the formulae ReP_3, ReP_2 and ReP.

REFERENCES

(1) Feit. *Z. angew. Chem.* 1931, **44**, 65.
(2) Briscoe, Robinson and Stoddart. *J. Chem. Soc.* 1931, p. 1439.
(3) Juza and Biltz. *Z. Elektrochem.* 1931, **37**, 498.
(4) Geilmann and Lange. *Z. anal. Chem.* 1944, **126**, 321.
Geilmann *et al.* *Z. anal. Chem.* 1944, **126**, 418.
(5) Noddack, W. and I. *Metallbörse,* 1931, **21**, 603 and 651.
(6) Becker and Moers. *Metallwirtschaft,* 1930, **9**, 1063.
(7) Haraldsen. *Z. anorg. Chem.* 1935, **221**, 398.

CHAPTER VII

SOME ORGANIC RHENIUM DERIVATIVES

CERTAIN organic derivatives have been prepared. One trialkyl rhenium compound has been obtained through the interaction of rhenium trichloride and Grignard's reagent, namely, trimethylrhenium, $Re(CH_3)_3$. This was described in 1934 (1), having been made by the interaction of an ethereal rhenium trichloride solution with magnesium methyl iodide,

$$2ReCl_3 + 6CH_3MgI = 2Re(CH_3)_3 + 3MgI_2 + 3MgCl_2.$$

Trimethylrhenium is an almost colourless, volatile oil of ethereal odour, boiling at about 60° C. It is somewhat heavier than water.

A similar oil, boiling-point about 80° C., was obtained by the interaction of rhenium trichloride and magnesium ethyl bromide. Its analysis agreed with that expected of triethylrhenium, $Re(C_2H_5)_3$. Like the trimethyl compound it was inflammable, although not especially so, giving off clouds of rhenium heptoxide.

Other attempts to prepare these rhenium alkyls were not successful (2) and have led to the view that possible undetectable impurities catalyse the decomposition of the products. Insufficient quantities of these organic derivatives have yet been obtained so that very little is known of their properties. They are not the main reaction products as a rule. A sharp odour, recalling that of certain tin alkyl halides, was very noticeable (3). A small amount of a halogen-containing rhenium compound was actually isolated and appeared to be di-methylrhenium chloride, $Re(CH_3)_2Cl$.

Rhenium oxythiocyanate, $ReO(CNS)_4$. When potassium thiocyanate is added to perrhenic acid, or potassium perrhenate in hydrochloric acid, in the presence of stannous chloride, a yellow colour develops and quickly becomes deep red. The formation of this rhenium oxythiocyanate may be expressed (4):

(*a*) $2HReO_4 + 10HCl + SnCl_2 = 2ReOCl_4 + SnCl_4 + 6H_2O,$

(*b*) $ReOCl_4 + 4KCNS = ReO(CNS)_4 + 4KCl.$

Rhenium oxythiocyanate is analogous to uranyl thiocyanate, $UO_2(CNS)_2$, $8H_2O$. These compounds resemble each other in being deliquescent and in readily dissolving in water and ether and other organic solvents. Both also formed double salts with pyridine and quinoline thiocyanates, these derivatives having the general formulae, $(B.HCNS)_2$, $ReO(CNS)_4$ and $(B.HCNS)_2$, $UO_2(CNS)_2$.

Rhenium oxythiocyanate, containing 41·7 % of the metal, can be isolated from the products of the interaction of perrhenate, thiocyanate and stannous chloride in hydrochloric acid by extraction with ether in which it is very soluble. As the solvent evaporates it leaves a dark red crystalline residue which can be dried in a desiccator and recrystallized from dry ether.

The substance softens but does not melt on heating; moreover, a yellow organic sublimate is also formed. When heated on platinum foil in air, rhenium oxythiocyanate burns with intumescence, leaving at first a dark residue which later ignites to form white fumes of rhenium heptoxide and the whole ultimately disappears.

The double thiocyanates of pyridine and quinoline separate as heavy oils when rhenium oxythiocyanate solution is added to pyridine or quinoline hydrochloride solutions with the appropriate amount of potassium thiocyanate. The oils contain water, but after remaining in a vacuum desiccator over magnesium perchlorate for some days they become anhydrous and solid.

Recently the reduction of perrhenates with stannous chloride and potassium thiocyanate has been reinvestigated by Tribalat (5) who has concluded that the stannous chloride reduces the perrhenate to quadrivalent rhenium quantitatively in the presence of four times normal hydrochloric acid. With stronger acid the reaction is not quantitative. Chromous chloride and titanous chloride can also be used, but they yield unstable derivatives of pentavalent rhenium.

With stannous chloride maximum coloration was obtained with normal hydrochloric acid and 2·5 times the theoretical amount of stannous chloride required by the equation.

According to Turkiewicz (6) complex double rhenium cyanides are formed when hydrated rhenium dioxide is acted upon by alkali cyanides. The compound obtained with potassium cyanide was $K_3[ReO(CN)_4 . OH]$, whilst in acid solutions potentiometric measurements were considered to prove the presence of a divalent ion, $[Re(CN)_4(OH)_2]''$. Pale pink crystals were deposited on adding alcohol to the reaction product from an aqueous solution of potassium perrhenate, potassium cyanide and hydrazine, and these were found to be $K_2NaReO(CN)_4$. The surprising presence of sodium is apparently to be attributed to some sodium salt being present in the cyanide used.

A green complex rhenium compound has also been made with toluene-3:4-dithiol (7). It is soluble in such organic solvents as *n*-butyl acetate or carbon tetrachloride. If the butyl acetate solution of rhenium oxythiocyanate is heated with hydrochloric acid and a little dithiol the colour deepens to red-brown in contrast to that developed by similar molybdenum and tungsten derivatives.

Rhenium triethoxide, $Re(OC_2H_5)_3$ and *tri*-iso-*propoxide*, $Re(OC_3H_7)_3$ are brown solids obtained by interaction of rhenium trichloride with sodium ethoxide or propoxide. They are stable in dry air but decompose with water, acids or alkalis and when heated alone (8).

Effect of Rhenium on Plant Growth

In recent years the significance of the small amounts of manganese or molybdenum (and some other elements) absorbed by plants has been receiving attention. It has been established that the ashes of most plants contain traces of manganese. The presence of rhenium has not yet been reported in plants growing normally, but it was considered of interest to study the effect of watering certain plants with potassium perrhenate solutions.

In some preliminary experiments cress seedlings were watered with a 0·1 % potassium perrhenate solution over a growing period of 3 weeks. They grew normally, that is, almost as well as the control seedlings watered with rain water. They grew better than some that were watered with a 0·1 % potassium permanganate solution.

Rhenium was unmistakably present in the ashes of those seedlings that had received the perrhenate solution.

When watered with similiar perrhenate solutions geraniums growing in pots tended to cast off leaves and these were found to contain rhenium. Specimens of mullein (*Verbascum Thapsus*) also did not do so well when treated with perrhenate solution, compared with untreated specimens.

THE TOXICITY OF RHENIUM

Some American investigators have examined (9) the physiological effects of administering potassium perrhenate to animals. Mice were given intraperitoneal injections in doses corresponding to 0·05–3 mg. of rhenium. All but one (accidental) recovered within 12 hr. The immediate effect was one of water intoxication on account of the low solubility of the salt.

Rats similarly injected with 2·5–50 mg. rhenium showed no unusual results when observed for a week, after which they were given 5 mg. more, and this was followed by nine daily injections of 2·5 mg. The physiological effects upon a dog were similarly negligible (very slight splanchnic dilatation).

In tracing the fate of the element in rabbits it was detected spectrographically half an hour after injection in the testes, heart, kidney, liver and spleen and especially in the urine.

REFERENCES

(1) Druce. *J. Chem. Soc.* 1934, p. 1129.

(2) Gilman, Jones, Moore and Kolbezen. *J. Amer. Chem. Soc.* 1941, **63**, 2525.

(3) Compare similar tin compounds, e.g. *J. Chem. Soc.* 1921, **119**, 758; 1922, **121**, 306 and *Rec. trav. chim. Pays-Bas*, 1925, **44**, 340.

(4) Druce. *Rec. trav. chim. Pays-Bas*, 1935, **54**, 334.

(5) Tribalat. *C.R. Acad. Sci., Paris*, 1946, **223**, 34.

(6) Turkiewicz. *Roczn. Chem.* 1932, **12**, 589.

(7) Miller. *J. Chem. Soc.* 1941, p. 792.

(8) Druce. *J. Chem. Soc.* 1937, pp. 1407–8.

(9) Hurd, Colehour and Cohen. *Proc. Soc. Exp. Biol., N.Y.*, 1933, **30**, 926.

Maresh, Lustok and Cohen. *Proc. Soc. Exp. Biol., N.Y.*, 1940, **45**, 576.

CHAPTER VIII

APPLICATIONS AND PATENTS RELATING TO RHENIUM

ATTEMPTS to find uses for rhenium and its compounds have led to the publication of a number of patents relating to the element. These fall broadly into two classes, namely, those dealing with its applications as a catalyst and those making use of certain properties of the element and its alloys for electro-technical and other purposes.

Both in the free state and when alloyed with copper, rhenium exhibits distinct catalytic properties. Thus, it resembles nickel in bringing about the hydrogenation of ethylene to ethane. In 1932 Tropsch and Kassler (1) examined its action as a catalyst in the conversion of carbon monoxide and hydrogen into methane,

$$2CO_2 + 5H_2 = 2H_2O + 2CH_4.$$

It was found that 4·8 % of the methane was formed at 350° C. and 18·7 % at 400° C., using pure rhenium as the catalyst. When the element was employed in conjunction with copper, these percentages rose to 14·3 and 31·8 respectively. Above 400° C. carbon began to deposit, and at 470° C. the decomposition of carbon monoxide was complete and the metals lost their catalytic powers, having been partly converted into carbides.

Rhenium, deposited on unglazed porcelain by reduction of material soaked in ammonium perrhenate solution in a stream of electrolytic hydrogen, has also been used as a catalyst in the dehydrogenation of propyl and isopropyl alcohols (2). The optimum temperature is again 400° C., whilst the yield is almost quantitative for both the formation of propaldehyde and for acetone.

Oxides of rhenium are also able to promote the formation of sulphur trioxide from the dioxide and oxygen, serving as oxygen carrier.

That rhenium has a future as a catalyst was the opinion expressed by Feit (3) in 1933. Its alloys with platinum and other noble metals might find use as thermocouples, and it is stated that one German concern has already used such a combination. The concern, Siemens and Halske, which holds the Noddacks' patent rights for the extraction of rhenium from molybdenum ores (4), have also patented a process for the manufacture of chloroform using rhenium as the catalytic agent (5). In these, rhenium and its alloys are claimed to be good catalysts in dehydrogenation processes.

Rhenium wire has been patented for use as a filament in electric-lamp bulbs and in wireless valves (6). It is said to volatilize even less than tungsten, but it is very doubtful if lamps or valves containing it have ever appeared on the market.

The coating of refractory metals (e.g. tungsten) with rhenium was covered in a specification in the name of the Treuhand Ges. für elektrischen Gluhlampen m.b.H. as long ago as 1931 (7). According to this the vapour of a volatile rhenium compound was conducted over tungsten wire by a current of inert and reducing gases (nitrogen and hydrogen) at a temperature of more than 2000° C. Even earlier claims for coating wires with rhenium were made. Thus, rhenium is mentioned in this connexion for coating wires, bars, threads, etc. (8), in 1931.

According to a French patent (9) unsaturated hydrocarbons with at least three carbon atoms to the molecule can be hydrogenated by rhenium or rhenium compounds at a sufficiently high temperature. Solid bodies can be impregnated with rhenium before hydrogenation as in the use of nickel for this purpose. The process is stated to be employed for making 'anti-knock' products.

Finally, alloys containing rhenium are claimed to be suitable for making pen-nibs (10). These alloys, with at least 2% of rhenium, are obtained with tungsten, chromium or tantalum (50–90%), the remainder being iron and (especially) nickel or cobalt up to 30%. An alternative is to alloy the rhenium with platinum.

Electroplating with rhenium is very readily accomplished although it is limited to thin coatings; these are very resistant to corrosive agents, even hydrochloric acid gas has no effect on them.

Thus, rhenium deposits can be used to protect such metals as silver. Bright, hard deposits are obtainable from both acid and alkali baths. The conditions have been studied by Fink and Deren (11), who used solutions of potassium perrhenate and also perrhenic acid itself. Other salts were added in various experiments, whilst the hydrogen-ion concentration was varied from

0·7 to 1·12 (with sulphate),
1·7 to 2·3 (with phosphate),
1·2 to 1·5 (with oxalate),
8·5 to 7·5 (with alkali).

Temperature was varied between 25 and 45° C., and current density was 10–14 amp./dm. All the baths possessed good 'throwing powers', and the action was complete in 1–1½ min.

Rhenium can also be co-deposited with other metals, and the disulphide, too, is a catalyst for the dehydrogenation of alcohols (12).

REFERENCES

(1) *Zprávy ústavu pro vědecké vyzkum uhli*, 1932, **2**, 13.
(2) Platonov, Anisimova and Kraseninnikova. *Ber. dtsch. chem. Ges.* 1936, **69**, 1050.
(3) *Angew. Chem.* 1933, **46**, 216.
(4) British Patent, 317,035, 28 July 1930; D.R.P., 483,495.
(5) British Patents, 346,652 and 358,180; Swiss Patent, 146,845; French Patent, 682,446 and D.R.P., 536,471.
(6) U.S. Patent, 1,829,756, 3 Nov. 1931.
(7) British Patent, 364,502, 4 June 1931.
(8) British Patent, 363,236, 27 Feb. 1931.
(9) French Patent, 761,632, 23 Mar. 1934.
(10) *Naturwissenschaften*, 1942, **30**, 149.
(11) *J. Amer. Electrochem. Soc.* 1934, **66**, 381.
(12) Platonov. *J. Gen. Chem.*, Russia, 1941, **11**, 683.

BIBLIOGRAPHY OF RHENIUM (DVI-MANGANESE)*

1925

W. Noddack, I. Tacke and O. Berg. Die Ekamangane. *Naturwissenschaften*, 1925, **13**, 567–74. Zwei neue Elemente der Mangangruppe. Chemischer Teil von W. Noddack and I. Tacke, Röntgenspektroskopischer Teil von O. Berg und I. Tacke. *S.B. preuss. Akad. Wiss.* 1925, **19**, 400–9.

J. G. F. Druce. Search for element 93. Part I. Examination of crude manganese compounds and the isolation of the element of atomic number 75. *Chem. News*, 1925, **131**, 273–7. Part II (with F. H. Loring). Examination of Crude Dvi-manganese. *Chem. News*, 1925, **131**, 337–8.

V. Dolejšek and J. Heyrovský. The occurrence of dvi-manganese (at. no. 75) in manganese salts. *Nature, Lond.*, 1925, **116**, 782; also *Bull. int. Acad. Sci. Bohême*, 1925, **25**, 179 and (in Czech) *Rozpravy České Akademie*, 1925, **34**, č. 25.

W. Noddack and I. Tacke. Zur Auffindung der Elemente Rhenium und Masurium. *Metallbörse*, 1925, **15**, 1597–9.

I. Tacke. Zur Auffindung der Ekamangane. *Z. angew. Chem.* 1925, **38**, 1157–60.

R. Swinne. Zwei neue Elemente, Masurium und Rhenium. *Z. techn. Phys.* 1925, **6**, 464–5.

O. Berg. Röntgenspektroskopie und die Nachweis der Ekamangane. *Z. techn. Phys.* 1925, **6**, 599–603.

A. N. Campbell. The occurrence of dvi-manganese in manganese salts. *Nature, Lond.*, 1925, **116**, 866.

1926

V. Dolejšek and J. Heyrovský. Zjistení přítomnosti dvimanganu (at. č. 75) v solech manganu. (Verification of the presence of dvi-manganese in manganese salts.) *Chem. Listy*, 1926, **20**, 4–12.

J. Heyrovský. The occurrence of dvi-manganese (at. no. 75) in manganese salts. *Nature, Lond.*, 1926, **117**, 16.

J. G. F. Druce. *Nature, Lond.*, 1926, **117**, 16.

F. H. Loring. Foreshadowing elements 75, 85, 87 and 93. *Nature, Lond.*, 1926, **117**, 153.

V. Dolejšek, J. G. F. Druce and J. Heyrovský. The occurrence of dvi-manganese in manganese salts. *Nature, Lond.*, 1926, **117**, 159.

* This chronological bibliography has been compiled with great care. Every effort has been made to include all papers relating to rhenium, including critical reviews. Twenty-one contributions of my own are recorded.

F. H. Loring. Dvi-manganese and eka-caesium. *Chem. News*, 1926, **132**, 101–2.

J. G. F. Druce. Dvi-manganese, the element of atomic number 75. *Sci. Progr.* 1926, **20**, 590–2.

J. G. F. Druce. The discovery of eka- and dvi-manganese. *Chem. Weekbl.* 1926, **23**, 318–20.

F. H. Loring. The problem of X-ray technique. *Nature, Lond.*, 1926, **117**, 622.

O. Zvjaginstsev. (Absence of) dvi-manganese in native platinum. *Nature, Lond.*, 1926, **117**, 262–3.

T. J. Patton and L. J. Waldbauer. Radioactivity of alkali metals. *Chem. Rev.* 1926, **3**, 81–93.

F. H. Loring. Notes on new elements. *Chem. News*, 1926, **132**, 407–10.

B. Polland. Limite d'absorption de la série K de l'élément de nombre atomique 75. *C.R. Acad. Sci., Paris*, 1926, **183**, 737–8.

W. F. Meggers and C. C. Kiess. Spectral structures for elements of the second long period. *J. Amer. Opt. Soc.* 1926, **12**, 417.

W. Prandtl. Auf der Suche nach den Manganhomologen Nr. 43 und 75. *Z. angew. Chem.* 1926, **39**, 1049–51.

W. and I. Noddack. Neuere Untersuchungen über die Elemente Masurium und Rhenium. *Metallbörse*, 1926, **16**, 2129–30.

W. and I. Noddack. Recent investigations of the elements masurium and rhenium. *Cont. Chem. Engng*, 1926, **1**, 109–11.

J. G. F. Druce. Dvi-manganese. *Cont. Chem. Engng*, 1926, **1**, 111–12.

J. G. F. Druce. Recent work on dvi-manganese. *Chem. Weekbl.* 1926, **23**, 497–8.

F. H. Loring. Search for the missing elements. *Chem. News*, 1926, **133**, 276–8 and (in German) *Z. angew. Chem.* 1927, **40**, 259–60.

F. H. Loring. On the synthesis of elements. *Chem. News*, 1926, **133**, 356–8.

I. and W. Noddack. Über die Elemente Masurium und Rhenium. Kritische Betrachtungen einiger neuen Untersuchungen. *Metallbörse*, 1926, **16**, 2633–4.

1927

J. G. F. Druce. Elements whose existence has been announced but which are not recognized. *Sci. Prog.* 1927, **21**, 479–91.

O. Zvjaginstsev, M. Korsunski and N. Seljakov. Dvi-mangan im Platinerz. *Z. angew. Chem.* 1927, **40**, 256–9.

N. Seljakov and M. Korsunski. Über den Nachweis der Ekamangane. *Phys. Z.* 1927, **28**, 478–9.

W. Prandtl. Zur Frage nach dem Vorkommen der Manganhomologen 43, 61 und 75. *Ber. dtsch. chem. Ges.* 1927, **60**, 621–3.

J. G. F. Druce. The interaction of manganese salts and permanganates. Part I. *Chem. News*, 1927, **134**, 145–7. Part II. *Chem. News*, 1927, **134**, 161–3.

G. von Hevesy. The use of X-rays for the discovery of new elements. *Chem. Rev.* 1927, **3**, 321–9.

V. Dolejšek and J. Heyrovský. Über die Vorkommen von Dvi-Mangan in Manganverbindungen. *Rev. trav. chim. Pays-Bas*, 1927, **46**, 248–55.

W. and I. Noddack. Über die Nachweis der Ekamangane. *Z. angew. Chem.* 1927, **40**, 250–4.

O. Berg. Röntgenspektroskopischer Nachweis des Rheniums. *Z. angew. Chem.* 1927, **40**, 254–6.

M. Herzfinkiel. Sur les élémentes de numèros atomiques 43, 61, 75, 85 et 87. *C.R. Acad. Sci., Paris*, 1927, **184**, 968.

I. and W. Noddack. Das Rhenium. *Ergebn. exakt. Naturw.* 1927, **6**, 333–73.

W. Prandtl. Auf der Suche nach den Manganhomologen Nr. 43 und 75. *Z. angew. Chem.* 1927, **40**, 561–4.

I. and W. Noddack. Darstellung und einige chemische Eigenschaften des Rheniums. *Z. phys. Chem.* 1927, **125**, 264–74.

O. Berg. Das Röntgenspektrum des Elements 75. *Phys. Z.* 1927, **28**, 864–6.

1928

H. Beuthe. Die *L*-Serie des Rheniums. *Z. Phys.* 1928, **46**, 873–7.

I. Wennerlöf. Präzisionsmessungen in der *L*-Serie des Rheniums. *Z. Phys.* 1928, **47**, 422–5.

O. Hahn. Neuere Fortschritte der Elementen- und Isotopen-Forschungen. *Z. angew. Chem.* 1928, **41**, 516.

E. Lindberg. Röntgenspektroskopische Messungen in der *M*-Serie der Elemente 66–92. *Z. Phys.* 1928, **50**, 82–96.

H. Beuthe. Weitere Röntgenspektroskopische Messungen in der *L*- und *M*-Serie des Rheniums. *Z. Phys.* 1928, **50**, 762–8.

W. Noddack. Beiträge zur Chemie des Rheniums. *Z. Elektrochem.* 1928, **34**, 627–9.

I. Noddack. Über einige physikalische Konstanten des Rheniums. *Z. Elektrochem.* 1928, **34**, 629–31.

1929

I. and W. Noddack. Die Sauerstoffverbindungen des Rheniums. *Naturwissenschaften*, 1929, **17**, 93–4.

V. M. Goldschmid. Die Krystallstruktur, Gitterkonstanten und Dichte des Rheniums. *Z. phys. Chem.* 1929, Abt. B, **2**, 244–52.

V. M. Goldschmid. Krystallstruktur des Rheniums. *Naturwissenschaften*, 1929, **17**, 134–5.

I. and W. Noddack. Die Sauerstoffverbindungen des Rheniums. *Z. anorg. Chem.* 1929, **181**, 1–37.

G. P. Baxter. Thirty-fifth Annual Report of the Committee on Atomic Weights. Determinations published in 1928. *J. Amer. Chem. Soc.* 1929, **51**, 647.

F. H. Loring. Mass numbers of the chemical elements and observations on element formation. *Chem. News*, 1929, **139**, 231.

E. Lindberg. Über die *M*-Serie des Rheniums. *Z. Phys.* 1929, **56**, 402–4.

E. Broch. Krystallstruktur des Kaliumperrhenat. *Z. phys. Chem.* 1929, Abt. B, **6**, 22–6.

I. and W. Noddack. Die Herstellung von einem Gramm Rheniums. *Z. anorg. Chem.* 1929, **183**, 353–75.

F. Schacherl. Dosavadní vysledky badaní o homologich manganu. (The present position of researches on manganese homologues.) *Chem. Listy*, 1929, **23**, 632–5.

W. and I. Noddack. Neuere Untersuchungen über die Eigenschaften des Rheniums. *Forsch. u. Fortschr.* 1929, **5**, 3.

C. E. St John. Elements unidentified or doubtful in the sun. *Astrophys. J.* 1929, **70**, 160–74.

1930

J. G. F. Druce. The chemistry of dvi-manganese. Element of at. no. 75. *Sci. Prog.* 1930, **24**, 480–5.

W. Feit. Über die technische Herstellung des Rheniums. *Z. angew. Chem.* 1930, **43**, 459–62.

I. and W. Noddack. Der Werdgang des Rheniums. *Metallbörse*, 1930, **20**, 621–2.

F. Machatschski. Krystallstruktur des Kaliumperrhenat. *Z. Krystallogr.* 1930, **72**, 541–2.

O. Hönigschmid and R. Sachtleben. Revision des Atomgewichtes des Rheniums. *Z. anorg. Chem.* 1930, **191**, 309–17.

K. Becker and K. Moers. Über die Schmelzpunkte im System Wolfram-Rheniums. *Metallwirtschaft*, 1930, **9**, 1063.

W. Geilmann and A. Voigt. Beiträge zur analytischen Chemie des Rheniums. I. Bestimmung löslicher Perrhenate mit Hilfe von Nitron. *Z. anorg. Chem.* 1930, **193**, 311–5.

F. Krauss and H. Steinfeld. Über die Darstellung von reinen Rheniumverbindungen. *Z. anorg. Chem.* 1930, **193**, 385–90.

N. A. Puškin and P. S. Tutundžič. Elektrolytische Leitfähigkeit der Lösungen von Kalium Perrhenat. *Z. anorg. Chem.* 1930, **193**, 420–4.

H. Tollert. Kaliumbestimmung als Kaliumperrhenat. *Naturwissenschaften*, 1930, **18**, 849.

R. A. Sonder. Über die Häufigkeitzahlen der Elemente und das Vorhandensein einer Kernperiodizität. *Naturwissenschaften*, 1930, **18**, 939–40.

H. Tropsch and R. Kassler. Über einige katalytische Eigenschaften des Rheniums. *Ber. dtsch. chem. Ges.* 1930, **63**, 2149–51.

A. Sandstrom. Röntgenspektroskopische Messungen des *L*-Absorption der Elementen 74–92. *Z. Phys.* 1930, **65**, 632–55.

1931

J. G. F. Druce. Atomic weight of rhenium. *Sci. Progr.* 1931, **25**, 499–500.

W. H. Albrecht and E. Wedekind. Magnetische Messungen am Rhenium. *Naturwissenschaften*, 1931, **19**, 20–1.

W. F. Meggers. The optical spectrum of rhenium. *Phys. Rev.* 1931, **37**, 219.

W. Feit. Schwefelderivaten der perrheniumsaure. *Z. angew. Chem.* 1931, **44**, 65–6.

W. Geilmann and F. Weibke. Die Bestimmung des Rheniums als Nitronperrhenat nach vorhergender Fällung als Sulfid. *Z. anorg. Chem.* 1931, **195**, 289–308.

J. G. F. Druce. The technical preparation of dvi-manganese (rhenium). *Industr. Chem.* 1931, **7**, 75.

C. Agte, H. Alterthum, K. Beckers, G. Heyne and K. Moers. Die physikalischen Eigenschaften des Rheniums. *Naturwissenschaften*, 1931, **19**, 108–9.

C. Agte, H. Alterthum, K. Becker, G. Heyne and K. Moers. Die physikalischen und chemischen Eigenschaften des Rheniums. *Z. anorg. Chem.* 1931, **196**, 129–59.

I. and W. Noddack. Zur Kenntniss des Rheniums. *Z. angew. Chem.* 1931, **44**, 215–20.

H. Schober and J. Birke. Die letzten Linien im Bogenspektrum des Rheniums. *Naturwissenschaften*, 1931, **19**, 211–12.

W. Meidinger. Messungen im Bogenspektrum des Rheniums. *Z. Phys.* 1931, **68**, 331–43.

F. Krauss and H. Steinfeld. Bestimmung des Rheniums als Thalloperrhenat. *Z. anorg. Chem.* 1931, **197**, 52–6.

W. F. Meggers. The spectrum of rhenium. *J. Franklin Inst.* 1931, **211**, 373–4.

H. Schober. Die Möglichkeit des Vorhandenseins von Rhenium unter den Frauenhoferschen Linien des Sonnenspektrums. *Naturwissenschaften*, 1931, **19**, 310.

H. V. A. Briscoe, P. L. Robinson and E. M. Stoddart. The reduction of potassium perrhenate. *J. Chem. Soc.* 1931, pp. 666–9.

I. and W. Noddack. Geochemie des Rheniums. *Z. phys. Chem.* 1931, Abt. A, **154**, 207–44.

F. Enk. Über rheniumchlorwasserstoffsaures Kalium. *Ber. dtsch. chem. Ges.* 1931, **64**, 791–7.

F. W. Aston. The constitution of rhenium. *Nature, Lond.*, 1931, **127**, 591.

F. H. Loring. The optical spectrum of rhenium and attempts to identify this element in the stars. *Chem. News*, 1931, **142**, 3212.

F. H. Loring. Resistivities of the metals at 18° C. and the problem of their relative abundance in the universe. *Chem. News*, 1931, **142**, 403–7.

E. Geay. Quelques élémentes rares. Hafnium, rhènium et indium. *Rev. Chim. industr.* 1931, **40**, 98–101.

R. Juza and W. Biltz. Über die Verwandschaft von Rhenium zu Schwefel. I. ReS_2. *Z. Elektrochem.* 1931, **37**, 498.

H. V. A. Briscoe, P. L. Robinson and E. M. Stoddart. The sulphides and selenides of rhenium. *J. Chem. Soc.* 1931, p. 1439.

K. Moeller. Zu den Gitterkonstanten des Rheniums. *Naturwissenschaften*, 1931, **19**, 575.

C. Agte and K. Becker. Rhenium. *Umschau*, 1931, **35**, 520–2.

W. Geilmann and F. W. Wrigge. Die Reaktionen des Rheniums auf trockenem Wege. *Z. anorg. Chem.* 1931, **199**, 65–76.

W. Geilmann and K. Brunger. Über einige mikrochemische Reaktionen des Rheniums. *Z. anorg. Chem.* 1931, **199**, 77–90.

W. Geilmann and F. Weibke. Die Abtrennung des Rheniums mit Chlorwasserstoff. *Z. anorg. Chem.* 1931, **199**, 120–8.

W. Feit. Über Monosulfperrheniumsaure. *Z. anorg. Chem.* 1931, **199**, 262–70.

W. Feit. Die spezifischen Gewichte wässrige Lösungen von Perrheniumsaure. *Z. anorg. Chem.* 1931, **199**, 271–3.

J. G. F. Druce. The preparation and properties of some compounds of rhenium. *Chem. News*, 1931, **143**, 66–7.

F. W. Aston. Isotopic constitution and atomic weight of...Re. *Proc. Roy. Soc.* A, 1931, **132**, 487–98.

H. V. A. Briscoe, P. L. Robinson and A. J. Rudge. The perrhenates of copper, nickel and cobalt. *J. Chem. Soc.* 1931, pp. 2211–13.

W. Geilmann and F. Weibke. Eine einfache Trennung von Molybdän und Rhenium. *Z. anorg. Chem.* 1931, **199**, 347–52.

N. A. Puškin and D. Kovač. Die Löslichkeit des Kaliumperrhenats in Wasser und einige physikalisch-chemische Konstante seiner Lösungen. *Z. anorg. Chem.* 1931, **199**, 369–73.

W. F. Meggers, A. S. King and R. F. Bacher. Hyperfine structure and nuclear moment of rhenium. *Phys. Rev.* 1931, **38**, 1258–9.

H. V. A. Briscoe, P. L. Robinson and E. M. Stoddart. Rhenium tetrachloride and the rhenichlorides. *J. Chem. Soc.* 1931, pp. 2263–8.

F. Krauss and H. Steinfeld. Zur Kenntniss der Verbindungen des 3- und 4-wertigen Rheniums. *Ber. dtsch. chem. Ges.* 1931, **64**, 2552–6.

M. LeBlanc and H. Sachse. Die Elektronenleitfähigkeit von festen Oxyden verschiedener Valenzstufen. *Phys. Z.* 1931, **32**, 887.

N. A. Puškin and D. Kovać. Rhenium and its compounds. *Bull. Soc. chim. Yugoslavie*, 1931, **2**, 111–28.

H. V. A. Briscoe, P. L. Robinson and A. J. Rudge. The alleged thallous thioperrhenate. *J. Chem. Soc.* 1931, pp. 2976–7.

H. V. A. Briscoe, P. L. Robinson and A. J. Rudge. A new oxide of rhenium. Rhenium pentoxide. *J. Chem. Soc.* 1931, pp. 3087–8.

W. F. Meggers. Rhenium. *Sci. Mon., N.Y.*, 1931, pp. 413–18 (Nov.).

H. Hölemann. Über einige Reduktionsprodukte, die bei der Elektrolyse von wässrigen Kaliumperrhenat Lösungen entstehen. *Z. anorg. Chem.* 1931, **202**, 277–91.

E. S. Kronman. The distribution of rhenium and its industrial production. *Mineralni Suir'e*, 1931, **6**, 945–55.

W. Biltz and G. A. Lehrer. Rheniumtrioxid. (m. Röntgenspektroskopischer Beiträge von K. Meisel. *Nachr. Ges. Wiss. Göttingen,* 1931, pp. 191–8.

I. and W. Noddack. Fortschritte in der Darstellung und Anwendung einiger seltenen Elemente. *Metallbörse,* 1931, **21**, 603–4 and 651–2.

W. Manchot, H. Schmid and J. Düsing. Über 3- u 4-wertiges Rhenium und sein Verhalten bei Oxydation. *Ber. dtsch. chem. Ges.* 1931, **64**, 2905–8.

P. Zeeman, J. H. Gisolf and T. L. de Bruin. Magnetic resolution and nuclear moment of rhenium. *Nature, Lond.,* 1931, **128**, 637.

O. Collenberg. Om Rhenium, dess upptackt och egenskaper. *Svensk kem. Tidskr.* 1931, **43**, 265–81.

L. A. Sommer and P. Karlson. Über das Kernmoment des Rheniums. *Naturwissenschaften,* 1931, **19**, 1021.

R. Schenk, F. Kurzen and H. Wesselkoch. Carbidstudien mit den Methanaufbaumethoden. *Z. anorg. Chem.* 1931, **203**, 159–87.

H. V. A. Briscoe, P. L. Robinson and A. J. Rudge. Potassium rheni-iodide. *J. Chem. Soc.* 1931, pp. 3218–20.

W. Biltz and F. Weibke. Über die Verwandschaft von Rhenium zu Schwefel. II. Re_2S_7. *Z. anorg. Chem.* 1931, **203**, 3–8.

1932

J. G. F. Druce. Further contributions to the chemistry of Re. *Chem. Weekbl.* 1932, **29**, 57–9.

H. Tollert. Die Bestimmung von Kalium mittels Perrheniumsaure. *Z. anorg. Chem.* 1932, **204**, 140–2.

W. A. Roth and G. Becker. Beiträge zur physikalischen Chemie des Rheniums. *Z. phys. Chem.* 1932, A, **159**, 27–39.

W. A. Roth and G. Becker. Rhenium Pentoxid. *Ber. dtsch. chem. Ges.* 1932, **65**, 373.

H. Tropsch and R. Kassler. O několikých Vlastnostech Rhenia jako Katalysatora. *Zpr. Úst. věd. Vyzk. Úhlí,* 1932, **2**, 13.

S. Piña de Rubies. Spectra of rhenium in analysis. *An. Soc. esp. Fís. Quím.* 1932, **30**, 918–21.

N. Perakis and L. Capatos. Sur le paramagnétisme constante du rhènium metallique. *C.R. Acad. Sci., Paris,* 1933, **196**, 611–2.

D. Vorlander, J. Hollatz and J. Fischer. Alkali Borfluoride, fluorsulfonate und Kaliumperrhenat. *Ber. dtsch. chem. Ges.* 1932, **65**, 535–8.

J. G. F. Druce. The thermite reaction with rhenium dioxide. *Chem. News,* 1932, **144**, 247.

H. V. A. Briscoe, P. L. Robinson and A. J. Rudge. The highest oxide of rhenium. *Nature, Lond.,* 1932, **129**, 618.

H. V. A. Briscoe, P. L. Robinson and A. J. Rudge. Rhenium oxychloride. *J. Chem. Soc.* 1932, pp. 1104–7.

F. Krauss and H. Dahlmann. Über Halogenverbindungen des Rheniums. *Ber. dtsch. chem. Ges.* 1932, **65**, 877–80.

A. Brukl and K. Ziegler. Rhenium oxychloride. *Ber. dtsch. chem. Ges.* 1932, **65**, 916–18.

E. Turkiewicz. Z Chemji nizszych Stopni Utelnienia Renu. *Rocz. Chem.* 1932, **12**, 589–97.

K. Meissel. Über die Krystallstruktur des Rheniumtrioxids. *Z. anorg. Chem.* 1932, **207**, 121–8.

J. G. F. Druce. Perrhenic acid and its salts. *Chem. & Ind.* 1932, pp. 632–3.

E. S. Kronman and E. Bibikova. Zum mikrochemischen Nachweis des Rheniums. *Mikrochemie*, 1932, **12**, 187–8.

E. Kronman. Zur analytischen Chemie des Rheniums. *Z. anal. Chem.* 1932, **90**, 31–4.

E. Kronman. Bemerkung zu der Arbeit 'Geochemie des Rheniums' von I. u. W. Noddack. *Z. phys. Chem.* 1932, A, **161**, 395–6.

E. S. Kronman. Renij. *J. Chem. Ind. Russia*, 1932, **145**, 186–7.

J. G. F. Druce. Volumetric estimation of potassium perrhenate. *Chem. News*, 1932, **145**, 186–7.

W. Geilmann, F. W. Wrigge and F. Weibke. Der Nachweis und Bestimmung kleiner Rheniummengen mit Hilfe von Kaliumrhodanid und Zinnchlorür. *Z. anorg. Chem.* 1932, **208**, 217–24.

H. Hagen and A. Sieverts. Über eine Farbreaktion zwischen Rheniumheptoxid und Wasserstoffperoxid und über Rheniumperoxid. *Z. anorg. Chem.* 1932, **208**, 367–8.

O. Ruff and W. Kwasnik (and E. Ascher). Die Fluorierung des Rheniums. *Z. anorg. Chem.* 1932, **209**, 113–22.

H. V. A. Briscoe, P. L. Robinson and A. J. Rudge. The parachor of rhenium. *J. Chem. Soc.* 1932, pp. 2673–6.

H. V. A. Briscoe, P. L. Robinson and E. M. Stoddart. The thioperrhenates. *J. Chem. Soc.* 1932, pp. 2811–12.

E. Einecke. Analytische Chemie und Physik des Rheniums. *Z. anal. Chem.* 1932, **90**, 127–9.

R. Dolique. Rhènium, élémente du numèro atomique 75. *Bull. Soc. pharmacol.* 1932, **39**, 677–86.

D. M. Yost and G. O. Shull. The density and molecular state of $ReCl_4$ and $ReCl_6$. *J. Amer. Chem. Soc.* 1932, **54**, 4657–61.

O. Michajlova, S. Pevsner and Archipova. Die Anwendung mikro-Methoden bei quantitätiver Bestimmung des Rheniums. *Z. anal. Chem.* 1932, **91**, 25–8.

E. Ogawa. Vapour pressure of rhenium heptoxide and dissociation pressure of Re_2O_7 and Re_2O_8. *Bull. Chem. Soc., Japan*, 1932, **7**, 265–73.

G. A. Aartovaara. (The occurrence of rare elements in Finland.) *Tek. Foren Finl. Forh.* 1932, **52**, 157–61.

W. Geilmann, F. W. Wrigge and W. Biltz. Rhenium Trichlorid. *Nachr. Ges. Wiss. Göttingen*, 1932, pp. 579–87.

1933

W. F. Jakob and B. Jeżowska. Über die Elektro-reduktion der Perrhensaure. *Ber. dtsch. chem. Ges.* 1933, **66**, 461–2.

W. Stenzel and J. Weerts. (Precision determinations of lattice constants of non-cubic substances.) *Z. Krystallogr.* 1933, **84**, 20–44.

W. Geilmann and L. C. Hurd. Die massanalytische Bestimmung der Rheniumoxyde. *Z. anorg. Chem.* 1933, **210**, 350–6.

W. Feit. Die technische Gewinnung des Rheniums und Galliums und einiger ihrer Verbindungen. *Angew. Chem.* 1933, **46**, 216–8.

N. Perakis and L. Capatos. Sur le paramagnétisme constant du Re metallique. *C.R. Acad. Sci., Paris*, 1933, **194**, 611–12.

E. Kronman and N. Berkmann. Neue mikrochemische Reaktionen des Rheniums. *Z. anorg. Chem.* 1933, **211**, 277–80.

W. Manchot and J. Düsing. Über die niedrigsten Wertigkeitsstufen von Rhenium und Ruthenium. *Z. anorg. Chem.* 1933, **212**, 21–31.

J. G. F. Druce. Some uses and properties of rhenium (dvi-manganese). *Industrial Chemist*, 1933, **9**, 244.

O. Ruff. Neues aus der Chemie des Fluor. *Angew. Chem.* 1933, **46**, 739.

H. Hölemann. Beiträge zur Chemie und Elektrochemie des Rheniums. II. *Z. anorg. Chem.* 1933, **211**, 195–203.

H. Schöber. Die Spektren des Rheniums. IV. *S.B. Akad. Wiss. Wien*, 1933, **142**, 35.

H. Schmid. Studien zur Chemie des Rheniums. I. Kalium-Rhenium iv-Chlorid und organische Derivate des 4-wertigen Rheniums. *Z. anorg. Chem.* 1933, **212**, 187–97.

J. H. Muller and W. A. Lalande. The precipitation of rhenium sulfide from ammoniacal solution. *J. Amer. Chem. Soc.* 1933, **55**, 2376–8.

W. Geilmann and L. C. Hurd. Die Bestimmung der Rhenichlorwasserstoffsaure neben Perrheniumsaure. *Z. anorg. Chem.* 1933, **213**, 336–42.

A. Schulze. (Thermo-elements at high temperatures.) *Z. Ver. dtsch. Ing.* 1933, **77**, 1241–2.

A. Brukl and E. Plettinger. Rheniumoxytetrachlorid. *Ber. dtsch. chem. Ges.* 1933, **66**, 971–3.

M. Prettre. Le Rhènium. *Bull. Soc. chim. Fr.* 1933, **53**, 669–81.

A. Englemann. Über die Bestimmung der lichtelektrischen Grenzwellenlänge an Rhenium. *Ann. Phys., Lpz.*, 1933, **17**, 185–208.

S. Piña de Rubies and J. Dorronsova. (Arc spectra of rhenium). *An. Soc. esp. Fís. Quím.* 1933, **31**, 412–15.

W. F. Meggers. Infra-red arc. Spectra of manganese and rhenium. *Bur. Stand. J. Res., Wash.*, 1933, **10**, 757–69.

F. M. Jaeger and J. Beintma. Crystal structure of perrhenates. *Proc. K. Akad. Wet. Amst.* 1933, **36**, 523–8.

F. M. Jaeger and E. Rosenbaum. The specific heat of metallic rhenium. *Proc. K. Akad. Wet. Amst.* 1933, **36**, 786–8.

E. Kronman, V. Bibikova and M. Aksenova. Über die Gewinnung von Rhenium aus Molybdänglanz. *Z. anorg. Chem.* 1933, **214**, 143–4.

L. C. Hurd. The discovery of rhenium. *J. Chem. Educ.* 1933, **10**, 605–8.

D. Vorländer and G. Dalichau. Schmelzpunkt und Siedpunkt des Kaliumperrhenats. *Ber. dtsch. chem. Ges.* 1933, **66**, 1534–6.

W. Biltz. Über Rheniumtrioxid und Rheniumdioxyd. *Z. anorg. Chem.* 1933, **214**, 225–38.

W. Geilmann and F. W. Wrigge. Über Rheniumsesquioxyd. *Z. anorg. Chem.* 1933, **214**, 239–43.

W. Geilmann, F.W. Wrigge and W. Biltz. Rheniumpentachlorid. *Z. anorg. Chem.* 1933, **214**, 244–7.

W. Geilmann and F. W. Wrigge. Über einige Reaktionen der Rheniumchloride. *Z. anorg. Chem.* 1933, **214**, 248–59.

W. Geilmann and L. C. Hurd. Über die Bestimmung des Rheniums als Dioxyd. *Z. anorg. Chem.* 1933, **214**, 260–8.

A. Brukl and K. Ziegler. Rheniumoxybromide. *Mh. Chem.* 1933, **63**, 329–34.

L. C. Hurd, J. L. Colehour and P. P. Cohen. Toxicity tests with potassium perrhenate. *Proc. Soc. Exp. Biol., N.Y.*, 1933, **30**, 926–8.

W. F. Jakob and B. Jeżowska. Über die elektrochemische Reduktion saurer Perrhenatlösungen. *Z. anorg. Chem.* 1933, **214**, 337–53.

H. Hagen and A. Sieverts. Rheniumtribromid. *Z. anorg. Chem.* 1933, **215**, 111–12.

I. and W. Noddack. Sauerstoff und Halogenverbindungen des Rheniums. *Z. anorg. Chem.* 1933, **215**, 129–84.

B. Scharnow. Über Mesoperrhenate. *Z. anorg. Chem.* 1933, **215**, 184–9.

E. Wilke-Dörfurt and T. Gunzert. Neue Perrheniumsauresalze. *Z. anorg. Chem.* 1933, **215**, 369–87.

W. Brandes and A. Geller. Die seltenen Elementen. *Z. prakt. Geol.* 1933, **41**, 153–63.

1934

J. G. F. Druce. The discovery of rhenium. *J. Chem. Educ.* 1934, **11**, 59.

H. Hölemann. Über die Reduktion der Perrhenate mit Zinn-2-chlorid und eine potentiometrische Bestimmungsmethode für siebenwertiges Rhenium. *Z. anorg. Chem.* 1934, **217**, 105–12.

B. Tougarinoff. Deux réactions de coloration du rhènium. *Bull. Soc. chim. Belg.* 1934, **43**, 111–14.

W. Manchot and J. Düsing. Über die unteren Wertigkeitsstufen des Rheniums. *Liebigs Ann.* 1934, **509**, 228–40.

J. G. F. Druce. Trimethylrhenium. *J. Chem. Soc.* 1934, p. 1129.

O. Ruff and W. Kwasník. Rheniumfluoride (insbesondere ReF_6, $ReOF_4$, ReF_4, ReO_2F_2u.K_2ReF_6). *Z. anorg. Chem.* 1934, **219**, 65–81.

O. Ruff and W. Kwasník. Rheniumfluoride. *Angew. Chem.* 1934, **47**, 480.

A. E. van Arkel. (Production of high-melting metals by thermal dissociation.) *Metallwirtschaft*, 1934, **13**, 405–8.

W. Biltz, W. Geilmann and F. W. Wrigge. Rheniumtrichlorid. *Liebigs Ann.* 1934, **511**, 301–3.

W. F. Jakob and B. Jeżowska. Über das fünfwertige Rhenium. *Z. anorg. Chem.* 1934, **220**, 16–20.

H. Hölemann. Über die Einwirkung von Reduktionsmitteln auf Schwefelsaure Lösungen von Kalium. *Z. anorg. Chem.* 1934, **220**, 33–40.

O. Ruff and W. Kwasník. Rheniumfluoride. *Z. anorg. Chem.* 1934, **220**, 96.

B. Jeżowska. (Reduction of perrhenic acid.) *Roczn. Chem.* 1934, **14**, 1061–87.

J. T. Dobbin and J. K. Colehour. The preparation of perrhenic acid. *J. Amer. Chem. Soc.* 1934, **56**, 2056.

C. H. Kao and T. L. Chang. Detection of rhenium in Noyes and Bray's system. *Chin. Chem. J.* 1934, **2**, 6–12.

V. I. Ivernova. (Precision methods for measuring parameters of crystal lattices.) *J. Tech. Phys.* (*U.S.S.R.*), 1934, **4**, 459–75.

N. Perakis, L. Kapitos and P. Kyriakides. Paramagnetism of rhenium. *Praktika*, 1934, **8**, 163–8.

C. G. Fink and P. Deren. Rhenium plating. Amer. Electrochem. Soc., read Sept. 1934; published *Bull. Amer. Electrochem. Soc.* **66**, 381–4.

W. Schuth and W. Klemm. (Magneto-chemical behaviour of some rhenium compounds.) *Z. anorg. Chem.* 1934, **220**, 193–8.

E. S. Kronman, V. I. Bibikova and M. Axenova. (Extraction of rhenium from molybdenite.) *J. Appl. Chem., Moscow*, 1934, **7**, 47–50.

J. A. Bearden and C. H. Shaw. Self-ionisation of Na and Cs at the glowing surfaces of Re and W. *Phys. Rev.* 1934, **46**, 754–63.

1935

H. Haraldsen. Das System Rhenium-Phosphor. *Z. anorg. Chem.* 1935, **221**, 398–417.

M. S. Platonov, G. B. Anisimov and V. M. Krasheninnikova. Über die katalytischen Eigenschaften des Rheniums. *Ber. dtsch. chem. Ges.* 1935, **68**, 761–5.

W. Geilmann and F. W. Wrigge. Über die massanalytische Bestimmungen der Wertigkeit von Rheniumverbindungen. *Z. anorg. Chem.* 1935, **222**, 56–64.

O. Hahn and L. Meitner. Über die künstliche Umwandlung des Urans durch Neutronen. *Naturwissenschaften*, 1935, **23**, 37–8.

J. G. F. Druce. Rhenium oxythiocyanate. *Rec. Trav. chim. Pays-Bas*, 1935, **54**, 334–5.

J. Heyrovský. A sensitive polarographic test for the absence of Re in Mn salts. *Nature, Lond.*, 1935, **135**, 870–1.

N. Perakis and L. Capatos. Magnetochimie du rhènium. *J. Phys. Radium*, 1935, **6**, 1462–8.

H. Anderssen. Das galvanische Überziehen mit Rhodium und Rhenium. *Chem. Z.* 1935, **59**, 375–6.

W. Geilmann and F. W. Wrigge. Über Doppelsalze von Rheniumtrichlorid mit Rubidium- und Cäsiumchlorid. *Z. anorg. Chem.* 1935, **223**, 144–8.

H. Lange. Die optischen Konstanten von Rhenium und Gallium für die Wellenlänge 589 u. 436 mμ. *Z. Phys.* 1935, **94**, 650–4.

1936

L. C. Hurd. Determination of rhenium. I. Qualitative. *Industr. Engng Chem.* (Anal. ed.), 1936, **8**, 11–15.

L. C. Hurd and B. J. Babler. The Geilmann reaction. *Industr. Engng Chem.* (Anal. ed.), 1936, **8**, 112–14.

H. Yahoda. Detection of rhenium with sodium carbonate bead. *Industr. Engng Chem.* (Anal. ed.), 1936, **8**, 133–4.

J. G. F. Druce. Precursors of rhenium. *Chem. & Ind.* 1936, **55**, 577–8.

S. A. Borovik and N. M. Gudris. (Last lines of rhenium in presence of molybdenum.) *J. Appl. Chem., Moscow,* 1936, **9**, 937–42.

C. B. F. Young. Plating rhenium and rhenium-nickel alloys. *Metal Ind., N.Y.,* 1936, **34**, 176–7.

I. P. Alimanin and B. N. Ivanov. (Chemical concentration of rhenium in determinations in oxide and sulphide ores.) *J. Appl. Chem., Moscow,* 1936, **9**, 1124–35.

M. S. Platonov, G. B. Anisimova and V. M. Krasheninnikova. Über die katalytischen Eigenschaften des Rheniums. II. Dehydrierung der Propylalkohole. *Ber. dtsch. chem. Ges.* 1936, **69**, 1050–3.

N. S. Poluektov. (Sensitive test for rhenium.) *J. Appl. Chem., Moscow,* 1936, **9**, 2312–14.

F. W. Wrigge and W. Biltz (with H. Brintzinger). Über den Molekularzustand von rotem Rheniumchlorid in Lösungen. *Z. anorg. Chem.* 1936, **228**, 372–82.

1937

W. Klemm and G. Frischmuth. Die Ammoniakate der Rheniumtrihalogenide. *Z. anorg. Chem.* 1937, **230**, 209–14.

W. Klemm and G. Frischmuth. Über ein komplexes Rheniumoxycyanid. *Z. anorg. Chem.* 1937, **230**, 215–19.

W. Klemm and G. Frischmuth. Weitere Untersuchungen an Rheniumverbindungen. *Z. anorg. Chem.* 1937, **230**, 220–2.

E. Neusser. Über Perrhenate einiger Kobaltammoniak-Komplexsalze. *Z. anorg. Chem.* 1937, **230**, 253–6.

W. Geilmann and F. W. Wrigge. Über einige Krystallreaktionen des Rheniumchlorids und der Rheniumchlorwasserstoffsäure. *Z. anorg. Chem.* 1937, **231**, 66–77.

M. S. Platonov and V. I. Tomilov. (Preparation of rhenium catalysts.) *J. Gen. Chem., Moscow,* 1937, **7**, 776–7.

M. S. Platonov and V. I. Tomilov. (Catalytic decomposition of formic acid and ethyl alcohol.) *J. Gen. Chem., Moscow,* 1937, **7**, 778–81.

M. S. Platonov and S. B. Anisimova. (Dehydration of butyl alcohols.) *J. Gen. Chem., Moscow*, 1937, **7**, 1360–3.

G. E. F. Lundell and H. B. Knowles. A contribution to the chemistry of rhenium. *Bur. Stand. J. Res., Wash.*, 1937, **18**, 629–37.

J. G. F. Druce. Ethoxides and iso-propoxides of manganese and rhenium. *J. Chem. Soc.* 1937, pp. 1407–8.

W. Trzebiatowski. Über die Beziehungen des Rheniums zum Kohlenstoff. *Z. anorg. Chem.* 1937, **233**, 376–84.

W. Biltz, F. W. Wrigge, E. Prange and G. Lange. Über die Bestimmung des Chlors in Rheniumverbindungen. *Z. anorg. Chem.* 1937, **234**, 142–9.

W. Biltz and G. Lange. Über die Bestimmung des Chlors in Rheniumverbindungen. *Z. anorg. Chem.* 1937, **234**, 289–97.

R. C. Young and J. W. Irvine. Preparation of two lower oxides of rhenium. *J. Amer. Chem. Soc.* 1937, **59**, 2648–50.

M. S. Platonov, V. I. Tomilov and E. V. Tur. (Decomposition of methyl alcohol with rhenium.) *J. Gen. Chem., Moscow*, 1937, **7**, 1803.

H. Hölemann. Untersuchungen über komplexe Rhodanide des Rheniums und über das Rhenium-(v)-oxychlorid. *Z. anorg. Chem.* 1937, **235**, 1–24.

C. Zenghelis and E. Stathis. Über die Ammoniaksynthese durch die katalytische Wirkung von metallischem Rhenium. *Öster. Chem. Z.* 1937, **40**, 80–1.

Y. Cauchois. Émissions faibles dans le spectre *L* du rhènium (75). *C.R. Acad. Sci., Paris*, 1937, **204**, 255–7.

Y. Cauchois. Les spectres d'émission et d'absorption du rhènium et ses nivaux caractéristiques. *J. Phys. Radium*, 1937, **8**, 267–72.

1938

H. Hölemann and W. Kleese. Die Löslichkeit des Kaliumperrhenats in Wasser zwischen 10 und 378° C. *Z. anorg. Chem.* 1938, **237**, 172–6.

J. G. F. Druce. Manganese perchlorate and perrhenate. *J. Chem. Soc.* p. 966.

J. G. F. Druce. Recent work on rhenium. *Sci. Progr.* 1938, **32**, 132–4.

Sir G. Morgan and Glynn R. Davies. Some complex compounds of rhenium. *J. Chem. Soc.* pp. 1858–61.

N. Perakis, T. Karantassis and L. Capatos. Atomic moment of quadrivalent rhenium. *C.R. Acad. Sci., Paris*, 1938, **206**, 1369–71.

F. Schmidt. (Magnetic moment of...rhenium....) *Z. Phys.* 1938, **108**, 408–20.

R. Fonteyne. Raman spectrum and structure of perrhenic acid and the perrhenate ion. *Natuurwet. Tijdschr.* 1938, **20**, 20–30.

L. C. Hurd and C. F. Hiskey. Determination of rhenium in pyrolusite. *Industr. Engng Chem.* (Anal. ed.). 1938, **10**, 623–6.

1939

E. Tiede and H. Lemke. (Preparation and detector action of the sulphides of tungsten, molybdenum and rhenium.) *Ber. dtsch. chem. Ges.* 1939, **71**, 582–8.

F. H. Loring. Some physico-chemical numerics. *Chem. Prod. and Chem. News*, 1939, **2**, 24–9.

B. Jeżowska-Trzebiatowska and C. Iodka. (Complex chlorides of quadrivalent rhenium). *Roczn. Chem.* 1939, **19**, 187.

O. and F. Tomíček. Electro-analytical determination of rhenium. *Trans. Electrochem. Soc.* 1939, **76**, 197–203.

H. Schulten. Über Rhenium-kohlenoxyd Verbindungen. *Z. anorg. Chem.* 1939, **243**, 164–73.

K. Sinma and F. Yamasaki. Beta-radioactivity of rhenium. *Phys. Rev.* 1939 (ii), **55**, 320.

F. Weichmann and M. Heimburg (with W. Biltz). (Affinity of rhenium for arsenic). *Z. anorg. Chem.* 1939, **240**, 129–38 (correction, p. 368).

1940

R. C. Young and P. M. Bernays. Possible source of error in determinations involving use of hydrogen peroxide followed by nitron. *Industr. Engng Chem.* (Anal. ed.), 1940, **12**, 90.

C. F. Hiskey and V. W. Meloche. Determination of rhenium in molybdenite minerals. *Industr. Engng. Chem.* (Anal. ed.), 1940, **12**, 503–6.

K. Fajans and W. H. Sullivan. Induced radioactivity of rhenium and tungsten. *Phys. Rev.* 1940, **58**, 276–8.

F. Maresh, M. J. Lustok and P. P. Cohen. Physiology of rhenium compounds. *Proc. Soc. Exp. Biol., N.Y.*, 1940, **45**, 576–9.

I. and W. Noddack and V. Bohnstedt. Distribution coefficients of heavy metals between FeS and Fe. *Z. anorg. Chem.* 1940, **244**, 252–80.

1941

H. Gilman, R. G. Jones, F. W. Moore and M. J. Kolbezen. Reaction of rhenium trichloride with magnesium methyl iodide. *J. Amer. Chem. Soc.* 1941, **63**, 2525–6.

L. P. Works. Rhenium-bearing molybdenite in northern Wisconsin. *Rocks and Minerals*, 1941, **16**, 92–3.

W. Hieber, R. Schuh and H. Fuchs. Preparation of rhenium carbonyl halides. *Z. anorg. Chem.* 1941, **248**, 243–76.

N. S. Poluektov. (Colorimetric determination of rhenium.) *J. Appl. Chem., Moscow*, 1941, **14**, 695–702.

N. S. Platonov. (Rhenium and its sulphide as catalysts.) *J. Gen. Chem., Moscow*, 1941, **11**, 683–5.

C. C. Miller. Qualitative semi-microanalysis with reference to Noyes and Bray's system. *J. Chem. Soc.* 1941, pp. 786–92.

C. C. Miller. Preliminary observations on the behaviour of rhenium and of the complex of rhenium and molybdenum with toluene-3:4-diol. *J. Chem. Soc.* 1941, p. 792.

1942

J. J. Lingane. Polarographic investigation of rhenium compounds. I. Reduction of perrhenate ions at the dropping mercury cathode. II. Oxidation of −1 Re to a potential of Re 2. *J. Amer. Chem. Soc.* 1942, **64**, 1001–7.

H. J. Wallbaum. Zirconium rhenide. *Naturwissenschaften*, 1942, **30**, 149.

H. J. Wallbaum. Rhenium silicide. *Metallkunde*, 1942, **33**, 378–81.

A. Voigt. (Analytical chemistry of rhenium.) *Z. anorg. Chem.* 1942, **249**, 225–8.

G. Aschermann and E. Justi. (Some physical properties of rhenium.) *Phys. Z.* 1942, **249**, 225–8.

P. Wenger and R. Duckert. (Reagents for rhenium cation and the perrhenate ion.) *Helv. Chim. Acta*, 1942, **25**, 207–12.

W. Hieber. (Rhenium.) *Die Chemie*, 1942, **55**, 7–11 and 24–8.

1943

J. G. F. Druce. Recent progress with rhenium. *Industrial Chemist*, 1943, **19**, 343–4.

V. V. Lebedinsky and B. N. Ivanov-Emin. (A new type of rhenium complex.) *J. Gen. Chem., Moscow*, 1943, **13**, 253–5.

O. Winkler. (Abrasion resistance). *Z. Elektrochem.* 1943, **49**, 221.

1944

W. Geilmann and G. Lange. (Precipitation and determination of rhenium as sulphide.) *Z. anal. Chem.* 1944, **126**, 321–34.

C. C. Miller. Toluene-3:4-dithiol as selective reagent for tungsten in presence of molybdenum and rhenium. *Analyst*, 1944, **69**, 112.

W. Geilmann (with F. Wiechmann, F. W. Wrigge *et al.*). (Precipitation of rhenium in various valencies as sulphide.) *Z. anal. Chem.* 1944, **126**, 418–26.

B. S. Anisimov. (Identification of rhenium by the drop method.) *J. Appl. Chem., Moscow*, 1944, **17**, 658–9.

1945

S. Tribalat. Penta- and hepta-valent rhenium. *C.R. Acad. Sci., Paris*. 1945, **220**, 885.

1946

S. Tribalat. Complex rhenium thiocyanate. *C.R. Acad. Sci., Paris*, 1946, **222**, 1388–90 and **223**, 34–6.

INDEX

Printed in the United States
By Bookmasters